TBILISI - MATHEMATICS

Tbilisi Mathematical Journal
Special Issues, 1

Editor Hvedri Inassaridze

DOI 10.2478/9788395793882-fm

Special issue on 8th International Eurasian Conference on Mathematical Sciences and Applications (IECMSA-2019), 27 – 30 August 2019, Baku, Azerbaijan
Lead Guest Editor: Murat Tosun (Sakarya University, Turkey)
Guest Editors: Cristina Flaut (Ovidius University, Romania),
Etibar S. Penahov (Baku State University, Azerbaijan),
H. Hidayet Kosal (Sakarya University, Turkey),
Lyudmila Romakina (Saratov State University),
Mehmet Ali Gungor (Sakarya University, Turkey),
Soley Ersoy (Sakarya University, Turkey),
Wolfgang Sproessig (Technical University Freiburg, Germany).

DOI 10.2478/9788395793882-001

New versions of q-surface pencil in Euclidean 3-space

Aziz Yazla[1] and Muhammed T. Sariaydin[2]

[1,2]Selcuk University, Faculty of Science, Department of Mathematics, 42130, Konya, Turkey
E-mail: azizyazla@gmail.com[1], talatsariaydin@gmail.com[2]

Abstract

In this paper, the q-surface pencil is studied in Euclidean 3-space. By using q-frame in Euclidean space, Firstly, we define q-surface pencil. Then, we give the necessary and sufficient condition for a curve to be a geodesic curve and to be an asymptotic curve on a q-surface pencil. Then, we study this subject for an offset q-surface pencil.

2010 Mathematics Subject Classification. **53A05** 53A04
Keywords. Geodesic Curve, Asymptotic Curve, q-Surface Pencil, Offset Surface.

1 Introduction

Geodesic on a surface corresponds to the shortest path between any two points on the surface. When one flattens a developable surface into a planar shape without distortion, any geodesic on the surface will be mapped to a straight line in the planar shape. Therefore, a good algorithm which is needed to flatten a non-developable surface with as little distortion as possible must preserve the geodesic curvatures on the surface, [16].

A geodesic on a surface is a parametric curve on the surface such that the acceleration vector on every point of the curve is orthogonal to the surface. It is well-known that the tangent vector field of a parametric curve on a surface is parallel in Levi-Civita sense if and only if this curve is a geodesic on the surface. A geodesic on a surface can also be defined as a curve on the surface with zero geodesic curvature, [10]. Additionally, there are many works related to surface pencil [1-3,8].

Geodesic curves play an important role in many industrial applications, such as tent manufacturing and textile manufacturing. Most existing work on geodesics is finding geodesics on a given surface. However, the reverse problem that is how to characterize the surfaces that possess a given curve as a common geodesic is also studied. Shoe design can be given as an engineering application of this problem. The characteristic curve of a woman shoe which is called the girth is wanted to be a geodesic on the shoe surface, [16].

One important concept in differential geometry of curves and surfaces is also asymptotic curves. An asymptotic curve on a surface is a curve such that its tangent vector field is an asymptotic direction. Along an asymptotic direction, the surface never makes twists from its tangent plane. Furthermore, the Gaussian curvature of the surface is never positive along an asymptotic direction. Moreover, the acceleration vector of the asymptotic curve on the surface is tangent to the surface, [4]. Additionally, there are many works related to this surfaces [11-14].

In this paper, we study q-surface pencil and its offset. Given a space curve with the q-frame, we obtain the conditions for this curve to be a common geodesic and to be a common asymptotic curve on a q-surface pencil. Then we give answers to the same problem for the offset q-surface pencil.

Tbilisi Mathematical Journal Special Issue (IECMSA-2019), pp. 1–10.
Tbilisi Centre for Mathematical Sciences.

Received by the editors: 20 September 2019.
Accepted for publication: 27 December 2019.

2 Preliminaries

In this section, we present the Frenet frame and the q-frame along a space curve. Also, we give some geometric properties for these frames.

Let $\alpha(s)$ be a space curve parameterized with arc-length in $\mathbb{R}^3$. The Frenet frame of $\alpha(s)$ is $\{\mathbf{T}(s), \mathbf{N}(s), \mathbf{B}(s)\}$, where the vector fields are given as

$$\begin{aligned} \mathbf{T}(s) &= \alpha'(s), \\ \mathbf{N}(s) &= \frac{\alpha''(s)}{\|\alpha''(s)\|}, \\ \mathbf{B}(s) &= \mathbf{T}(s)\times\mathbf{N}(s). \end{aligned}$$

Here, $\mathbf{T}(s)$ is called the tangent vector field, $\mathbf{N}(s)$ is called the principal normal vector field and $\mathbf{B}(s)$ is called the binormal vector field of the curve $\alpha(s)$.

The curvature $\kappa(s)$ and the torsion $\tau(s)$ of the curve $\alpha(s)$ are given by

$$\begin{aligned} \kappa(s) &= \|\alpha''(s)\|, \\ \tau(s) &= \frac{\det(\alpha'(s), \alpha''(s), \alpha'''(s))}{\|\alpha''(s)\|^2}. \end{aligned}$$

The q-frame of a space curve $\alpha(s)$ which is parameterized with arc-length is $\{\mathbf{t}_q(s), \mathbf{n}_q(s), \mathbf{b}_q(s)\}$, where the vector fields are given as

$$\begin{aligned} \mathbf{t}_q(s) &= \mathbf{T}(s), \\ \mathbf{n}_q(s) &= \frac{\mathbf{T}(s)\times\overrightarrow{\mathbf{k}}}{\left\|\mathbf{T}(s)\times\overrightarrow{\mathbf{k}}\right\|}, \\ \mathbf{b}_q(s) &= \mathbf{T}(s)\times\mathbf{n}_q(s). \end{aligned}$$

Here, $\overrightarrow{\mathbf{k}}$ is the projection vector which can be chosen as $\overrightarrow{\mathbf{k}} = (1,0,0)$ or $\overrightarrow{\mathbf{k}} = (0,1,0)$ or $\overrightarrow{\mathbf{k}} = (0,0,1)$. In this paper, we choose the projection vector $\overrightarrow{\mathbf{k}} = (0,0,1)$. $\mathbf{n}_q(s)$ and $\mathbf{b}_q(s)$ are called the quasi normal vector field and the quasi binormal vector field of the curve $\alpha(s)$, respectively, [5].

Let $\theta(s)$ be the angle between the principal normal vector field $\mathbf{N}(s)$ and the quasi normal vector field $\mathbf{n}_q(s)$. The quasi formulas are given by

$$\frac{d}{ds}\begin{bmatrix} \mathbf{t}_q(s) \\ \mathbf{n}_q(s) \\ \mathbf{b}_q(s) \end{bmatrix} = \begin{bmatrix} 0 & k_1(s) & k_2(s) \\ -k_1(s) & 0 & k_3(s) \\ -k_2(s) & -k_3(s) & 0 \end{bmatrix}\begin{bmatrix} \mathbf{t}_q(s) \\ \mathbf{n}_q(s) \\ \mathbf{b}_q(s) \end{bmatrix},$$

where $k_i(s)$ are called the quasi curvatures $(1 \le i \le 3)$ which are given by

$$\begin{aligned} k_1(s) &= \kappa(s)\cos\theta(s) = \left\langle \mathbf{t}_q'(s), \mathbf{n}_q(s)\right\rangle, \\ k_2(s) &= -\kappa(s)\sin\theta(s) = \left\langle \mathbf{t}_q'(s), \mathbf{b}_q(s)\right\rangle, \\ k_3(s) &= \theta'(s) + \tau(s) = -\left\langle \mathbf{n}_q(s), \mathbf{b}_q'(s)\right\rangle. \end{aligned}$$

The relationship between the Frenet frame and the q-frame is given by, [6]. Additionally, there are many works related to q-frame, [7,15].

$$\begin{bmatrix} \mathbf{T}(s) \\ \mathbf{N}(s) \\ \mathbf{B}(s) \end{bmatrix} = \begin{bmatrix} 1 & 0 & 0 \\ 0 & \cos\theta(s) & -\sin\theta(s) \\ 0 & \sin\theta(s) & \cos\theta(s) \end{bmatrix} \begin{bmatrix} \mathbf{t}_q(s) \\ \mathbf{n}_q(s) \\ \mathbf{b}_q(s) \end{bmatrix}.$$

3 q-Surface pencil with a common geodesic curve

We can define q-surface pencil similar to the definition of surface pencil in [9], as follows:

Definition 3.1. q-surface pencil possessing $\alpha(s)$ as a common curve is given by

$$S(s,t) = \alpha(s) + x_1(s,t)\mathbf{t}_q(s) + x_2(s,t)\mathbf{n}_q(s) + x_3(s,t)\mathbf{b}_q(s), \tag{3.1}$$

$0 \leq s \leq L$, $0 \leq t \leq T$, where $x_1(s,t)$, $x_2(s,t)$ and $x_3(s,t)$ are C^2 functions. The curve $\alpha(s)$ is a common parametric curve on the q-surface pencil $S(s,t)$ if

$$x_1(s,t_0) = x_2(s,t_0) = x_3(s,t_0) = 0.$$

Theorem 3.2. Let $\alpha(s)$ be a parametric curve on the q-surface pencil $S(s,t)$ which is given as (3.1). Then, $\alpha(s)$ is a geodesic curve on the surface $S(s,t)$ if and only if

$$\mu_1(s,t_0) = 0,\ \mu_2(s,t_0) = k\cos\theta(s),\ \mu_3(s,t_0) = -k\sin\theta(s), \tag{3.2}$$

where $m(s,t_0) = \mu_1(s,t_0)\mathbf{t}_q(s) + \mu_2(s,t_0)\mathbf{n}_q(s) + \mu_3(s,t_0)\mathbf{b}_q(s)$ is the normal vector field of $S(s,t)$ along the curve $\alpha(s)$, $\theta(s)$ is the angle between the principal normal vector field $\mathbf{N}(s)$ and the quasi normal vector field $\mathbf{n}_q(s)$ of the curve $\alpha(s)$ and $k \in \mathbb{R}$, $k \neq 0$.

Proof. The normal vector field of $S(s,t)$ is

$$\begin{aligned} m(s,t) &= \frac{\partial S}{\partial s}(s,t) \times \frac{\partial S}{\partial t}(s,t) \\ &= \mu_1(s,t)\mathbf{t}_q(s) + \mu_2(s,t)\mathbf{n}_q(s) + \mu_3(s,t)\mathbf{b}_q(s), \end{aligned}$$

where

$$\begin{aligned} \mu_1(s,t) &= (\frac{\partial x_3}{\partial t}(s,t)k_1(s) - \frac{\partial x_2}{\partial t}(s,t)k_2(s))x_1(s,t) \\ &\quad -(\frac{\partial x_3}{\partial t}(s,t)x_3(s,t) + \frac{\partial x_2}{\partial t}(s,t)x_2(s,t))k_3(s) \\ &\quad +\frac{\partial x_3}{\partial t}(s,t)\frac{\partial x_2}{\partial s}(s,t) - \frac{\partial x_2}{\partial t}(s,t)\frac{\partial x_3}{\partial s}(s,t), \\ \mu_2(s,t) &= (\frac{\partial x_1}{\partial t}(s,t)k_3(s) + \frac{\partial x_3}{\partial t}(s,t)k_1(s))x_2(s,t) \\ &\quad +(\frac{\partial x_3}{\partial t}(s,t)x_3(s,t) + \frac{\partial x_1}{\partial t}(s,t)x_1(s,t))k_2(s) \\ &\quad +\frac{\partial x_1}{\partial t}(s,t)\frac{\partial x_3}{\partial s}(s,t) - \frac{\partial x_3}{\partial t}(s,t)\frac{\partial x_1}{\partial s}(s,t) - \frac{\partial x_3}{\partial t}(s,t), \end{aligned}$$

$$\begin{aligned}\mu_3(s,t) &= (\frac{\partial x_1}{\partial t}(s,t)k_3(s) - \frac{\partial x_2}{\partial t}(s,t)k_2(s))x_3(s,t)\\ &-(\frac{\partial x_2}{\partial t}(s,t)x_2(s,t) + \frac{\partial x_1}{\partial t}(s,t)x_1(s,t))k_1(s)\\ &+\frac{\partial x_2}{\partial t}(s,t)\frac{\partial x_1}{\partial s}(s,t) - \frac{\partial x_1}{\partial t}(s,t)\frac{\partial x_2}{\partial s}(s,t) + \frac{\partial x_2}{\partial t}(s,t).\end{aligned}$$

Along the curve $\alpha(s)$, we can write

$$m(s,t_0) = \mu_1(s,t_0)\mathbf{t}_q(s) + \mu_2(s,t_0)\mathbf{n}_q(s) + \mu_3(s,t_0)\mathbf{b}_q(s),$$

where

$$\begin{aligned}\mu_1(s,t_0) &= \frac{\partial x_3}{\partial t}(s,t_0)\frac{\partial x_2}{\partial s}(s,t_0) - \frac{\partial x_2}{\partial t}(s,t_0)\frac{\partial x_3}{\partial s}(s,t_0),\\ \mu_2(s,t_0) &= -(1+\frac{\partial x_1}{\partial s}(s,t_0))\frac{\partial x_3}{\partial t}(s,t_0) - \frac{\partial x_3}{\partial t}(s,t_0)\frac{\partial x_1}{\partial s}(s,t_0),\\ \mu_3(s,t_0) &= (1+\frac{\partial x_1}{\partial s}(s,t_0))\frac{\partial x_2}{\partial t}(s,t_0) - \frac{\partial x_1}{\partial t}(s,t_0)\frac{\partial x_2}{\partial s}(s,t_0).\end{aligned}$$

Since $\alpha(s)$ is a parametric curve on the q-surface pencil $S(s,t)$, we can write

$$x_1(s,t_0) = x_2(s,t_0) = x_3(s,t_0) = 0.$$

We know that $\alpha(s)$ is a geodesic curve on the surface $S(s,t)$ if and only if the normal vector field $m(s,t_0)$ of the surface $S(s,t)$ along the curve $\alpha(s)$ is parallel to the principal normal vector field $\mathbf{N}(s)$ of the curve $\alpha(s)$, that is, $m(s,t_0)\|\mathbf{N}(s)$, [16]. Using the relationship between the Frenet frame and the q-frame, we obtain

$$\begin{aligned}m(s,t_0)\|\mathbf{N}(s) &\Leftrightarrow m(s,t_0)\|(\cos\theta(s)\mathbf{n}_q(s) - \sin\theta(s)\mathbf{b}_q(s))\\ &\Leftrightarrow m(s,t_0) = k\cos\theta(s)\mathbf{n}_q(s) - k\sin\theta(s)\mathbf{b}_q(s)\\ &\Leftrightarrow \mu_1(s,t_0) = 0,\ \mu_2(s,t_0) = k\cos\theta(s),\ \mu_3(s,t_0) = -k\sin\theta(s).\end{aligned}$$

Q.E.D.

Corollary 3.3. Let $\alpha(s)$ be a parametric curve on the q-surface pencil $S(s,t)$ which is given as (3.1). Then, $\alpha(s)$ is a geodesic curve on the surface $S(s,t)$ if and only if

$$(\frac{\partial x_2}{\partial t}(s,t_0))^2 + (\frac{\partial x_3}{\partial t}(s,t_0))^2 = k^2, \tag{3.3}$$

where $k \in \mathbb{R}$, $k \neq 0$.

Proof. Since $\alpha(s)$ is a parametric curve on the q-surface pencil $S(s,t)$, we can write

$$x_1(s,t_0) = x_2(s,t_0) = x_3(s,t_0) = 0.$$

With the help of the definition of partial differentiation, we get

$$\frac{\partial x_1}{\partial s}(s,t_0) = \frac{\partial x_2}{\partial s}(s,t_0) = \frac{\partial x_3}{\partial s}(s,t_0) = 0.$$

So, we get $\mu_1(s,t_0)=0$, $\mu_2(s,t_0)=-\frac{\partial x_3}{\partial t}(s,t_0)$ and $\mu_3(s,t_0)=\frac{\partial x_2}{\partial t}(s,t_0)$. From Theorem 3.2, we know that $\alpha(s)$ is a geodesic curve on the surface $S(s,t)$ if and only if

$$\mu_1(s,t_0)=0,\ \mu_2(s,t_0)=k\cos\theta(s),\ \mu_3(s,t_0)=-k\sin\theta(s),$$

where $k\in\mathbb{R}$, $k\neq 0$. Therefore, we obtain $\alpha(s)$ is a geodesic curve on the surface $S(s,t)$ if and only if

$$(\frac{\partial x_2}{\partial t}(s,t_0))^2+(\frac{\partial x_3}{\partial t}(s,t_0))^2=k^2.$$

Q.E.D.

Theorem 3.4. Let $S(s,t)$ be a q-surface pencil which is given as (3.1), where

$$\begin{aligned} x_1(s,t) &= \varphi_1(s)\psi_1(t),\\ x_2(s,t) &= \varphi_2(s)\psi_2(t),\\ x_3(s,t) &= \varphi_3(s)\psi_3(t), \end{aligned}$$

$0\le s\le L$, $0\le t\le T$, $\varphi_1(s)$, $\varphi_2(s)$, $\varphi_3(s)$, $\psi_1(t)$, $\psi_2(t)$, $\psi_3(t)$ are C^1 functions and $\varphi_1(s)$, $\varphi_2(s)$, $\varphi_3(s)$ are not identically zero. Then, $\alpha(s)$ is a parametric and a geodesic curve on the surface $S(s,t)$ if and only if

$$\psi_1(t_0)=\psi_2(t_0)=\psi_3(t_0)=0, \tag{3.4}$$

$$(\varphi_3(s)\psi_3'(t_0))^2+(\varphi_2(s)\psi_2'(t_0))^2=k^2, \tag{3.5}$$

where $k\in\mathbb{R}$, $k\neq 0$.

Proof. We know that $\alpha(s)$ is a parametric curve on the surface $S(s,t)$ if and only if

$$x_1(s,t_0)=x_2(s,t_0)=x_3(s,t_0)=0.$$

Since the functions $\varphi_1(s)$, $\varphi_2(s)$, $\varphi_3(s)$ are not identically zero, we get $\alpha(s)$ is a parametric curve on the surface $S(s,t)$ if and only if

$$\psi_1(t_0)=\psi_2(t_0)=\psi_3(t_0)=0.$$

Assuming $\alpha(s)$ to be a parametric curve on the surface $S(s,t)$, simple calculations give us

$$\mu_2(s,t_0)=-\varphi_3(s)\psi_3'(t_0) \text{ and } \mu_3(s,t_0)=\varphi_2(s)\psi_2'(t_0).$$

From Corollary 3.3, we obtain $\alpha(s)$ is a geodesic curve on the surface $S(s,t)$ if and only if

$$(\varphi_3(s)\psi_3'(t_0))^2+(\varphi_2(s)\psi_2'(t_0))^2=k^2,$$

where $k\in\mathbb{R}$, $k\neq 0$. Q.E.D.

Theorem 3.5. Let $S(s,t)$ be a q-surface pencil which is given as (3.1), where

$$\begin{aligned} x_1(s,t) &= \varphi_1(s)+\psi_1(t),\\ x_2(s,t) &= \varphi_2(s)+\psi_2(t),\\ x_3(s,t) &= \varphi_3(s)+\psi_3(t), \end{aligned}$$

$0 \leq s \leq L$, $0 \leq t \leq T$, $\varphi_1(s)$, $\varphi_2(s)$, $\varphi_3(s)$, $\psi_1(t)$, $\psi_2(t)$, $\psi_3(t)$ are C^1 functions and $\varphi_1(s)$, $\varphi_2(s)$, $\varphi_3(s)$ are not identically zero. Then, $\alpha(s)$ is a parametric and a geodesic curve on the surface $S(s,t)$ if and only if

$$\psi_1(t_0) = \psi_2(t_0) = \psi_3(t_0) = 0, \tag{3.6}$$

$$(\psi_2'(t_0))^2 + (\psi_3'(t_0))^2 = k^2, \tag{3.7}$$

where $k \in \mathbb{R}$, $k \neq 0$.

Proof. We know that $\alpha(s)$ is a parametric curve on the surface $S(s,t)$ if and only if

$$x_1(s,t_0) = x_2(s,t_0) = x_3(s,t_0) = 0.$$

Since the functions $\varphi_1(s)$, $\varphi_2(s)$, $\varphi_3(s)$ are not identically zero, we get $\alpha(s)$ is a parametric curve on the surface $S(s,t)$ if and only if

$$\begin{aligned} \varphi_1(s) &= -\psi_1(t_0) \neq 0, \\ \varphi_2(s) &= -\psi_2(t_0) \neq 0, \\ \varphi_3(s) &= -\psi_3(t_0) \neq 0. \end{aligned}$$

Assuming $\alpha(s)$ to be a parametric curve on the surface $S(s,t)$, simple calculations give us

$$\mu_2(s,t_0) = -\psi_3'(t_0) \text{ and } \mu_3(s,t_0) = \psi_2'(t_0).$$

From Corollary 3.3, we obtain $\alpha(s)$ is a geodesic curve on the surface $S(s,t)$ if and only if

$$(\psi_2'(t_0))^2 + (\psi_3'(t_0))^2 = k^2,$$

where $k \in \mathbb{R}$, $k \neq 0$. Q.E.D.

4 q-Surface pencil with a common asymptotic curve

Theorem 4.1. Let $S(s,t)$ be a q-surface pencil which is given as (3.1) and $\alpha(s)$ be a parametric curve on the surface $S(s,t)$. Then, $\alpha(s)$ is an asymptotic curve on the surface $S(s,t)$ if and only if

$$k_1(s)\frac{\partial x_3}{\partial t}(s,t_0) = k_2(s)\frac{\partial x_2}{\partial t}(s,t_0), \tag{4.1}$$

where $k_1(s)$, $k_2(s)$ are the quasi curvatures of the curve $\alpha(s)$.

Proof. We know that a parametric curve $\alpha(s)$ on a surface is an asymptotic curve if and only if the normal vector field $m(s,t_0)$ of the surface along the curve $\alpha(s)$ is perpendicular to the principal normal vector field $\mathbf{N}(s)$ of the curve $\alpha(s)$, that is, $\langle m(s,t_0), \mathbf{N}(s)\rangle = 0$. Also, we know that $\langle m(s,t_0), \mathbf{T}(s)\rangle = 0$, [4]. By differentiating both sides of this equation, one can get

$$\left\langle \frac{\partial m}{\partial s}(s,t_0), \mathbf{T}(s) \right\rangle = 0. \tag{4.2}$$

So, this equality becomes the necessary and sufficient conditon for $\alpha(s)$ to be an asymptotic curve on the surface. The normal vector field $m(s,t_0)$ of the surface $S(s,t)$ along the curve $\alpha(s)$ is

$$m(s,t_0) = \mu_1(s,t_0)\mathbf{t}_q(s) + \mu_2(s,t_0)\mathbf{n}_q(s) + \mu_3(s,t_0)\mathbf{b}_q(s), \tag{4.3}$$

where

$$\begin{aligned}\mu_1(s,t_0) &= \frac{\partial x_3}{\partial t}(s,t_0)\frac{\partial x_2}{\partial s}(s,t_0) - \frac{\partial x_2}{\partial t}(s,t_0)\frac{\partial x_3}{\partial s}(s,t_0),\\ \mu_2(s,t_0) &= -(1+\frac{\partial x_1}{\partial s}(s,t_0))\frac{\partial x_3}{\partial t}(s,t_0) - \frac{\partial x_3}{\partial t}(s,t_0)\frac{\partial x_1}{\partial s}(s,t_0),\\ \mu_3(s,t_0) &= (1+\frac{\partial x_1}{\partial s}(s,t_0))\frac{\partial x_2}{\partial t}(s,t_0) - \frac{\partial x_1}{\partial t}(s,t_0)\frac{\partial x_2}{\partial s}(s,t_0).\end{aligned}$$

Since $\alpha(s)$ is a parametric curve on the surface $S(s,t)$,

$$x_1(s,t_0) = x_2(s,t_0) = x_3(s,t_0) = 0.$$

Using this fact, we get

$$\begin{aligned}\mu_1(s,t_0) &= 0,\\ \mu_2(s,t_0) &= -\frac{\partial x_3}{\partial t}(s,t_0),\\ \mu_3(s,t_0) &= \frac{\partial x_2}{\partial t}(s,t_0),\end{aligned}$$

so we find

$$m(s,t_0) = -\frac{\partial x_3}{\partial t}(s,t_0)\mathbf{n}_q(s) + \frac{\partial x_2}{\partial t}(s,t_0)\mathbf{b}_q(s). \tag{4.4}$$

Then, one can easily get

$$\begin{aligned}\frac{\partial m}{\partial s}(s,t_0) &= (k_1(s)\frac{\partial x_3}{\partial t}(s,t_0) - k_2(s)\frac{\partial x_2}{\partial t}(s,t_0))\mathbf{t}_q(s)\\ &\quad -(\frac{\partial^2 x_3}{\partial s\partial t}(s,t_0) + k_3(s)\frac{\partial x_2}{\partial t}(s,t_0))\mathbf{n}_q(s)\\ &\quad +(\frac{\partial^2 x_2}{\partial s\partial t}(s,t_0) - k_3(s)\frac{\partial x_3}{\partial t}(s,t_0))\mathbf{b}_q(s).\end{aligned}$$

From condition (4.2), we derive

$$\begin{aligned}\left\langle \frac{\partial m}{\partial s}(s,t_0), \mathbf{T}(s)\right\rangle &= 0 \Leftrightarrow \left\langle \frac{\partial m}{\partial s}(s,t_0), \mathbf{t}_q(s)\right\rangle = 0\\ &\Leftrightarrow k_1(s)\frac{\partial x_3}{\partial t}(s,t_0) - k_2(s)\frac{\partial x_2}{\partial t}(s,t_0) = 0\\ &\Leftrightarrow k_1(s)\frac{\partial x_3}{\partial t}(s,t_0) = k_2(s)\frac{\partial x_2}{\partial t}(s,t_0).\end{aligned}$$

Same result can also be obtained as follows:

$$\begin{aligned}\langle m(s,t_0), \mathbf{N}(s)\rangle &= 0 \Leftrightarrow \langle m(s,t_0), k_1(s)\mathbf{n}_q(s) + k_2(s)\mathbf{b}_q(s)\rangle = 0\\ &\Leftrightarrow k_1(s)\langle m(s,t_0), \mathbf{n}_q(s)\rangle + k_2(s)\langle m(s,t_0), \mathbf{b}_q(s)\rangle = 0\\ &\Leftrightarrow -k_1(s)\frac{\partial x_3}{\partial t}(s,t_0) + k_2(s)\frac{\partial x_2}{\partial t}(s,t_0) = 0\\ &\Leftrightarrow k_1(s)\frac{\partial x_3}{\partial t}(s,t_0) = k_2(s)\frac{\partial x_2}{\partial t}(s,t_0).\end{aligned}$$

Consequently, $\alpha(s)$ is an asymptotic curve on the surface $S(s,t)$ if and only if

$$k_1(s)\frac{\partial x_3}{\partial t}(s,t_0) = k_2(s)\frac{\partial x_2}{\partial t}(s,t_0).$$

Q.E.D.

5 Offset q-surface pencil with a common geodesic and a common asymptotic curve

We can define offset q-surface pencil similar to the definition of offset surface pencil in [9], as follows:

Definition 5.1. Let $S(s,t)$ be a q-surface pencil which is given as (3.1), $m(s,t)$ be its unit normal vector field and $\alpha(s)$ be a parametric curve on the surface $S(s,t)$. The surface defined by

$$\overline{S}(s,t) = S(s,t) + dm(s,t) \tag{5.1}$$

is called an offset q-surface pencil, where $d \in \mathbb{R}$, $d \neq 0$.

Theorem 5.2. Let $\overline{S}(s,t) = S(s,t) + dm(s,t)$ be an offset q-surface pencil. The parametric curve $\beta(s) = \alpha(s) + dm(s,t_0)$ on $\overline{S}(s,t)$ is a geodesic if and only if

$$\begin{aligned}
&\frac{\partial^2 x_3}{\partial s \partial t}(s,t_0)(k_2(s)k_3(s) + 2k_1(s)) + \frac{\partial^2 x_2}{\partial s \partial t}(s,t_0)(k_1(s)k_3(s) \\
&\quad -2k_2(s)) + \frac{\partial x_3}{\partial t}(s,t_0)k_1'(s) - \frac{\partial x_2}{\partial t}(s,t_0)k_2'(s) = 0, \\
&\frac{\partial^2}{\partial s^2}(\frac{\partial x_3}{\partial t}(s,t_0)) + 2k_3(s)\frac{\partial^2 x_2}{\partial s \partial t}(s,t_0) + (k_1(s)k_2(s) + k_3'(s)) \\
&\quad \times\frac{\partial x_2}{\partial t}(s,t_0) - ((k_1(s))^2 + (k_3(s))^2 + \frac{c}{d})\frac{\partial x_3}{\partial t}(s,t_0) = \frac{k_1(s)}{d}, \\
&\frac{\partial^2}{\partial s^2}(\frac{\partial x_2}{\partial t}(s,t_0)) - 2k_3(s)\frac{\partial^2 x_3}{\partial s \partial t}(s,t_0) + (k_1(s)k_2(s) + k_3'(s)) \\
&\quad \times\frac{\partial x_3}{\partial t}(s,t_0) - (2(k_2(s))^2 + \frac{c}{d})\frac{\partial x_2}{\partial t}(s,t_0) = -\frac{k_2(s)}{d},
\end{aligned} \tag{5.2}$$

where $k_1(s)$, $k_2(s)$, $k_3(s)$ are the quasi curvatures of the curve $\alpha(s)$ and $c \in \mathbb{R}$, $c \neq 0$.

Proof. Let $\overline{m}(s,t)$ be the normal vector field of $\overline{S}(s,t)$ and $\overline{\mathbf{N}}(s)$ be the principal normal vector field of the curve $\beta(s)$. Then, $\beta(s)$ is a geodesic curve on the surface $\overline{S}(s,t)$ if and only if the normal vector field $\overline{m}(s,t_0)$ of the surface $\overline{S}(s,t)$ along the curve $\beta(s)$ is parallel to $\overline{\mathbf{N}}(s)$, that is, $\overline{m}(s,t_0)\|\overline{\mathbf{N}}(s)$. Since $\overline{m}(s,t_0) = \pm m(s,t_0)$ and $\overline{\mathbf{N}}(s) = \frac{\beta''(s)}{\|\beta''(s)\|}$, we get $\beta(s)$ is a geodesic curve on the surface $\overline{S}(s,t)$ if and only if $m(s,t_0)\|\beta''(s)$. We know that

$$m(s,t_0) = -\frac{\partial x_3}{\partial t}(s,t_0)\mathbf{n}_q(s) + \frac{\partial x_2}{\partial t}(s,t_0)\mathbf{b}_q(s),$$

So, the necessary and sufficient condition for $\beta(s)$ to be a geodesic curve on the surface $\overline{S}(s,t)$ becomes

$$\beta''(s) = -c\frac{\partial x_3}{\partial t}(s,t_0)\mathbf{n}_q(s) + c\frac{\partial x_2}{\partial t}(s,t_0)\mathbf{b}_q(s),$$

where $c \in \mathbb{R}$, $c \neq 0$. One can easily see that

$$\beta''(s) = \alpha''(s) + d\frac{\partial^2 m}{\partial s^2}(s,t_0). \tag{5.3}$$

Simple calculations give us

$$\alpha''(s) = k_1(s)\mathbf{n}_q(s) + k_2(s)\mathbf{b}_q(s),$$

and

$$\begin{aligned}\frac{\partial^2 m}{\partial s^2}(s,t_0) &= \left[\begin{array}{c}\frac{\partial^2 x_3}{\partial s\partial t}(s,t_0)(k_2(s)k_3(s)+2k_1(s)) + \frac{\partial^2 x_2}{\partial s\partial t}(s,t_0)(k_1(s)k_3(s) \\ -2k_2(s)) + \frac{\partial x_3}{\partial t}(s,t_0)k_1'(s) - \frac{\partial x_2}{\partial t}(s,t_0)k_2'(s)\end{array}\right]\mathbf{t}_q(s) \\ &- \left[\begin{array}{c}\frac{\partial^2}{\partial s^2}(\frac{\partial x_3}{\partial t}(s,t_0)) + 2k_3(s)\frac{\partial^2 x_2}{\partial s\partial t}(s,t_0) + (k_1(s)k_2(s)+k_3'(s)) \\ \times\frac{\partial x_2}{\partial t}(s,t_0) - ((k_1(s))^2+(k_3(s))^2)\frac{\partial x_3}{\partial t}(s,t_0)\end{array}\right]\mathbf{n}_q(s) \\ &+ \left[\begin{array}{c}\frac{\partial^2}{\partial s^2}(\frac{\partial x_2}{\partial t}(s,t_0)) - 2k_3(s)\frac{\partial^2 x_3}{\partial s\partial t}(s,t_0) + (k_1(s)k_2(s)+k_3'(s)) \\ \times\frac{\partial x_3}{\partial t}(s,t_0) - 2(k_2(s))^2\frac{\partial x_2}{\partial t}(s,t_0)\end{array}\right]\mathbf{b}_q(s).\end{aligned}$$

Putting these values in equality (5.3), we obtain the necessary and sufficient condition for $\beta(s)$ to be a geodesic curve on the surface $\overline{S}(s,t)$ as indicated in the theorem. Q.E.D.

Theorem 5.3. Let $\overline{S}(s,t) = S(s,t) + dm(s,t)$ be an offset q-surface pencil. The parametric curve $\beta(s) = \alpha(s) + dm(s,t_0)$ on $\overline{S}(s,t)$ is an asymptotic curve if and only if

$$\begin{aligned}&\left\{k_2(s) + d\left[\begin{array}{c}\frac{\partial^2}{\partial s^2}(\frac{\partial x_2}{\partial t}(s,t_0)) - 2k_3(s)\frac{\partial^2 x_3}{\partial s\partial t}(s,t_0) + (k_1(s)k_2(s)+k_3'(s)) \\ \times\frac{\partial x_3}{\partial t}(s,t_0) - 2(k_2(s))^2\frac{\partial x_2}{\partial t}(s,t_0)\end{array}\right]\right\}\frac{\partial x_2}{\partial t}(s,t_0) \\ = &\left\{k_1(s) - d\left[\begin{array}{c}\frac{\partial^2}{\partial s^2}(\frac{\partial x_3}{\partial t}(s,t_0)) + 2k_3(s)\frac{\partial^2 x_2}{\partial s\partial t}(s,t_0) + (k_1(s)k_2(s)+k_3'(s)) \\ \times\frac{\partial x_2}{\partial t}(s,t_0) - ((k_1(s))^2+(k_3(s))^2)\frac{\partial x_3}{\partial t}(s,t_0)\end{array}\right]\right\}\frac{\partial x_3}{\partial t}(s,t_0),\end{aligned}$$

where $k_1(s)$, $k_2(s)$ and $k_3(s)$ are the quasi curvatures of the curve $\alpha(s)$.

6 Conclusion

Surface pencil with a common geodesic curve is studied by Wang et al. in [16] and surface pencil with a common asymptotic curve is studied by Bayram et al. in [4]. They used the Frenet frame of the curves in their studies. In this work, we consider this subject using the q-frame of the curves. Moreover, we study offset q-surface pencil and give the necessary and sufficient conditions for a parametric curve on an offset q-surface pencil to be a geodesic curve and to be an asymptotic curve on the surface.

Acknowledgements

In this study, the authors were supported by the Selcuk University Scientific Research Project Fund (BAP). Support Fund under contract number 19701026.

References

[1] G.S. Atalay, E. Bayram, E. Kasap, *Surface Family with A Common Asymptotic Curve in Minkowski 3-Space*, Journal of Science and Arts, **18(2)** (2018) 357-368.

[2] G.S. Atalay, E. Kasap, *Surfaces Family With Common Smarandache Geodesic Curve*, Journal of Science and Arts, **17(4)** (2017) 651-664.

[3] E. Bayram, M. Bilici, *Surface family with a common involute asymptotic curve*, International Journal of Geometric Methods in Modern Physics, **13(5)** (2016) 1650062.

[4] E. Bayram, F. Guler, E. Kasap, *Parametric representation of a surface pencil with a common asymptotic curve*, Computer-Aided Design, **44(7)** (2012) 637-643.

[5] M. Dede, C. Ekici, A. Gorgulu, *Directional q-frame along a space curve*, Int Jour Adv Res Comp Scie Soft Eng IJARCSSE **5(12)** (2015) 775-780.

[6] M. Dede, C. Ekici,H. Tozak, *Directional tubular surfaces*, International Journal of Algebra, **9(12)** (2015) 527-535.

[7] C. Ekici, M.B. Göksel, M. Dede, *Smarandache Curves According to q-Frame in Minkowski 3-Space*, In Conference Proceedings of Science and Technology, **2(2)** (2019) 110-118.

[8] E. Ergün, E. Bayram, *Surface Family with a Common Natural Geodesic Lift*, Mathematical Combinatorics, **1** (2016) 34-41.

[9] F. Guler, E. Bayram, E. Kasap, *Offset surface pencil with a common asymptotic curve*, International Journal of Geometric Methods in Modern Physics **15** (2018).

[10] E. Kasap, F.T. Akyildiz, K. Orbay, *A generalization of surfaces family with common spatial geodesic*, Applied Mathematics and Computation, **201(1-2)** (2008) 781-789.

[11] T. Körpınar, V. Asil, Y. Ünlütürk, *An Approach for on* Π_1-*Surfaces of Biharmonic Constant* Π_2-*Slope Curves According to Type-2 Bishop Frame in The Sol Space*, In 17th International Geometry Symposium, Erzincan/Turkey, **117** (2019).

[12] T. Körpinar, E Turhan, *A new version of time-pencil surfaces around Biharmonic particles and its Lorentz transformations in Heisenberg spacetime*, International Journal of Theoretical Physics, **53(7)** (2014) 2288-2303.

[13] T. Körpınar, M.T. Sariaydin, *New Representation Of The Surface Pencil According To The Modified Orthogonal Frame With Curvature In Euclidean 3 Space*, Advanced Modeling And Optimization, **19(2)** (2017).

[14] M.T. Sariaydin, T. Korpinar, *On D-Pencil Surface By Using Darboux Frame In Minkowski 3-Space*, Journal of Science and Arts, **19(2)** (2019) 281-294.

[15] H. Tozak, C. Ekici, M. Dede, *Spacelike Translation Surface According to Q-Frame in Minkowski 3-Space*, Mathematical Studies and Applications, **4-6** (2018) 373.

[16] G.J. Wang, K. Tang, C.L. Tai, *Parametric representation of a surface pencil with a common spatial geodesic*, Computer-Aided Design, **36(5)** (2004) 447-459.

DOI 10.2478/9788395793882-002

A study on spacelike rectifying curves in Minkowski 3-space

Bahar Doğan Yazici[1] and Sıddıka Özkaldı Karakuş[2]

[1,2]Department of Mathematics, Bilecik Seyh Edebali University, 11000 Bilecik, Turkey
[1]Corresponding Author
E-mail: bahar.dogan@bilecik.edu.tr[1], siddika.karakus@bilecik.edu.tr[2]

Abstract

In this paper, we investigate spacelike rectifying curves via dilation of unit speed spacelike or timelike curves on Lorentzian unit spheres in Minkowski 3-space E_1^3. Then, we define a the centrode $D_\alpha(s)$ of a unit speed spacelike curve $\alpha(s)$ in E_1^3. In last section, we prove that if a unit speed spacelike curve $\alpha(s)$ in E_1^3 is neither a planar spacelike curve nor a helix, then its dilated centrode $\beta(s) = \rho_\alpha(s)D_\alpha(s)$, with dilation factor $\rho_\alpha(s)$, is always a rectifying curve, where ρ_α is the radius of curvature of α.

2010 Mathematics Subject Classification. **51B20** 53A35
Keywords. spacelike rectifying curve, centrode, dilated centrode, Minkowski 3-space.

1 Introduction

Let $\alpha : I \to E_1^3$ be a unit speed spacelike curve in Minkowski 3-space E_1^3 with Frenet-Serret apparatus $\{\kappa_\alpha, \tau_\alpha, T_\alpha, N_\alpha, B_\alpha\}$ where $\kappa_\alpha, \tau_\alpha, T_\alpha, N_\alpha$ and B_α denote the curvature, the torsion, the unit tangent T_α, the unit principal normal N_α and the unit binormal B_α of α, respectively.

Some important types of curves are helices (characterized by $\tau_\alpha = c\kappa_\alpha$ with a nonzero constant c), spherical curves (characterized by $(\rho'_\alpha \sigma_\alpha)' + \dfrac{\rho_\alpha}{\sigma_\alpha} = 0$ with $\rho_\alpha = \kappa_\alpha^{-1}$=radius of curvature, $\sigma_\alpha = \tau_\alpha^{-1}$=radius of torsion) and finally, rectifying curves given by $\dfrac{\tau_\alpha}{\kappa_\alpha} = as + b$ with constants $a \neq 0, b$.

The notion of rectifying curves was introduced by B.Y.Chen in [2]. By definition, a regular unit speed space curve $\alpha(s)$ in E^3 is called a rectifying curve, if its position vector always lies in its rectifying plane.

In [7], some characterizations of rectifying curves given by Ilarslan and Nesovic in Euclidean space. Also, Ilarslan, Nesovic and Petrovic-Torgasev have investigated rectifying curves in Minkowski space [9].

As spacelike rectifying curves are important, so is the relation between the Frenet-Serret apparatus $\{\kappa_\alpha, \tau_\alpha, T_\alpha, N_\alpha, B_\alpha\}$ of the spacelike rectifying curve $\alpha(t) = f(t)y(t)$ and that of the unit speed non-null curve $y(t)$. In this paper, we derive the Frenet-Serret apparatus of the spacelike rectifying curve $\alpha(t)$ in terms of that of the unit speed non-null curve $y(t)$.

Moreover, it is known that centrodes (i.e angular velocity vectors) play some important roles in mechanics and joint kinematics [1, 6, 15, 17, 18]. Regarding the centrode $D_\alpha = \tau_\alpha T_\alpha - \kappa_\alpha B_\alpha$ of a unit speed spacelike curve in E_1^3, it was shown in [8] that the centrode of a unit speed spacelike curve $\alpha : I \to E_1^3$ with non-zero constant curvature κ_α and non-constant torsion τ_α is a spacelike rectifying curve and vice versa.

In [4], rectifying curves as centrodes and extremal curves in Euclidean space are studied by Chen and Dillen. After them Ilarslan and Nesovic studied rectifying curves as centrodes and extremal

Tbilisi Mathematical Journal Special Issue (IECMSA-2019), pp. 11–26.
Tbilisi Centre for Mathematical Sciences.

Received by the editors: 25 October 2019.
Accepted for publication: 06 January 2020.

curves in Minkowski 3-space [8]. In [19], extended rectifying curves in Minkowski 3-space are studied by Yılmaz, Gök and Yaylı.

In this paper, we study the spacelike rectifying curves in Minkowski 3-space. By using similar methods as in [5] we study spacelike rectifying curve as centrode and dilated centrode.

2 Preliminaries

The Minkowski 3-space E_1^3 is Euclidean 3-space E^3 provided with the standard flat metric given by

$$g = -dx_1^2 + dx_2^2 + dx_3^2,$$

where (x_1, x_2, x_3) is a rectangular coordinate system of E_1^3. g is defined that a vector $v \in E_1^3$ can be spacelike if $g(v,v) > 0$ or $v = 0$, timelike if $g(v,v) < 0$ and null if $g(v,v) = 0$ and $v \neq 0$. Moreover, the norm(length) of a vector v is given by $\|v\| = \sqrt{|g(v,v)|}$, two vectors v and w are said to be orthogonal, if $g(v,w) = 0$. An arbitrary curve $\alpha(s)$ in E_1^3, can locally be spacelike, timelike or null, if all its velocity vectors $\alpha^{'}(s)$ are respectively spacelike, timelike or null. If $g(\alpha^{'}(s), \alpha^{'}(s)) = \pm 1$, the non-null curve α is said to be of unit speed (or parameterized by arc-length function s).

The Frenet frame $\{T_\alpha, N_\alpha, B_\alpha\}$ of a unit speed spacelike curve $\alpha(s)$ in E_1^3, with $g(\alpha^{''}(s), \alpha^{''}(s)) \neq 0$ for each s, is given by $T_\alpha(s) = \alpha^{'}(s), N_\alpha(s) = \dfrac{\alpha^{''}(s)}{\|\alpha^{''}(s)\|}, B_\alpha(s) = T_\alpha(s) \times N_\alpha(s)$. Let us put $g(T_\alpha, T_\alpha) = 1$ and $g(N_\alpha, N_\alpha) = \varepsilon = \pm 1$. Then $g(B_\alpha, B_\alpha) = -\varepsilon$ and the following Frenet formulas hold [12]:

$$\begin{aligned} T_\alpha^{'}(s) &= \kappa_\alpha(s) N_\alpha(s), \\ N_\alpha^{'}(s) &= -\varepsilon\kappa_\alpha(s) T_\alpha(s) + \tau_\alpha(s) B_\alpha(s), \\ B_\alpha^{'}(s) &= \tau_\alpha(s) N_\alpha(s). \end{aligned} \tag{2.1}$$

Accordingly, the Frenet frame of α satisfies the equations,

$$\begin{aligned} T_\alpha \times N_\alpha &= B_\alpha, \\ N_\alpha \times B_\alpha &= -\varepsilon T_\alpha, \\ B_\alpha \times T_\alpha &= -N_\alpha. \end{aligned} \tag{2.2}$$

We also recall that the pseudosphere of radius 1 and center at the origin is the hyperquadric in E_1^3 defined by

$$S_1^2 = \{v \in E_1^3 : g(v,v) = 1\},$$

and the pseudohyperbolic space of radius 1 and center at the origin is the hyperquadric in E_1^3 defined by

$$H_0^2 = \{v \in E_1^3 : g(v,v) = -1\}.$$

Let $\alpha = I \to E_1^3$ be a unit speed spacelike curve with curvature $\kappa_\alpha \neq 0$ and let $\{\kappa_\alpha, \tau_\alpha, T_\alpha, N_\alpha, B_\alpha\}$ be the Frenet-Serret apparatus of α. The distance function $f(s) = \|\alpha(s)\|$ of the spacelike rectifying curve satisfies

$$f(s) = \sqrt{s^2 + c_1 s + c_2},$$

where c_1 and c_2 are constants and the converse is also true. Moreover, it is also known that the unit speed spacelike curve α is a spacelike rectifying curve if and only if the ratio of torsion τ_α and curvature κ_α satisfies,

$$\frac{\tau_\alpha}{\kappa_\alpha} = as + b,$$

for some constants $a \neq 0$ and b [7].

The centrode of $\alpha : I \to E_1^3$ is defined by,

$$D_\alpha = \tau_\alpha T_\alpha - \kappa_\alpha B_\alpha,$$

which is the angular velocity vector of the motion of a mass particle along the spacelike curve α and it obeys the laws of motion:

$$\begin{aligned} T_\alpha^{'} &= D_\alpha \times T_\alpha, \\ N_\alpha^{'} &= D_\alpha \times N_\alpha, \\ B_\alpha^{'} &= D_\alpha \times B_\alpha. \end{aligned} \tag{2.3}$$

We shall find the curvature κ_y of the unit speed spacelike curve $y(t)$, which will be used subsequent work in this paper. Note that $T_y = y^{'}$ and that $\{y, y^{'}, y \times y^{'}\}$ is an orthonormal frame of E_1^3 and thus using Frenet-Serret formulae for $y(t)$ and

$$y^{''} = y + hy \times y^{'}, \tag{2.4}$$

with $h = g(y^{''}, y \times y^{'})$.

From (2.4) we have

$$T_y = y^{'}, \quad N_y = (\frac{1}{\kappa_y} y + \frac{h}{\kappa_y} y \times y^{'}). \tag{2.5}$$

It follows from the second equation in (2.5) that

$$\kappa_y = \sqrt{|h^2 - 1|}. \tag{2.6}$$

3 Some important results

In this section we recall some theorems from [8, 9], which are important for the proofs of theorems which follow.

Theorem 3.1. Let $\alpha = \alpha(s)$ be a unit speed non-null curve in E_1^3. Then the following statements hold:

(i) α is a rectifying curve with a spacelike rectifying plane if and only if, up to the parameterization, α is given by

$$\alpha(t) = \frac{a}{\cos(t)} y(t), a \in R_0^+,$$

where $y(t)$ is a unit speed spacelike curve lying in S_1^2.
(ii) α is a spacelike (resp. timelike) rectifying curve with a timelike rectifying plane and a spacelike (resp. timelike) position vector, if and only if up to the parameterization, α is given by

$$\alpha(t) = \frac{a}{\sinh(t)} y(t), a \in R_0^+,$$

where $y(t)$ is a unit speed timelike (resp. spacelike) curve lying in $S_1^2 (resp. H_0^2)$.
(iii) α is a spacelike (resp. timelike) rectifying curve with a timelike rectifying plane and a timelike (resp. spacelike) position vector, if and only if up to the parameterization, α is given by

$$\alpha(t) = \frac{a}{\cosh(t)} y(t), a \in R_0^+,$$

where $y(t)$ is a unit speed spacelike (resp. timelike) curve lying in $H_0^2 (resp. S_1^2)$ [9]

Theorem 3.2. The centrode of a unit speed spacelike curve $\alpha(s)$ in E_1^3, with constant curvature $\kappa_\alpha \neq 0$, non-constant torsion and $g(\alpha''(s), \alpha''(s)) \neq 0$ is a spacelike rectifying curve. Conversely, every unit speed spacelike rectifying curve in E_1^3, is the centrode of some unit speed spacelike curve with constant curvature $\kappa_\alpha \neq 0$ and non-constant torsion [8].

4 Spacelike rectifying curves via dilation of spacelike or timelike curves on Lorentzian unit spheres

In this section, firstly, we assume that α is a spacelike rectifying curve with a timelike position vector and $\alpha(t) = \frac{a}{\cosh(t+t_0)} y(t), a > 0, t_0 \in \mathbb{R}$ where $y(t)$ unit speed spacelike curve lying in $H_0^2 \subset E_1^3$ centered at the origin. However, if we consider an arc of the great circle, $y(t) = (\cosh t, 0, \sinh t)$ and the spacelike curve,

$$\begin{aligned} \alpha(t) &= \frac{a}{\cosh(t+t_0)} y(t), \\ &= a\left(\frac{1}{\cosh(t+t_0)} \cosh(t), 0, \frac{1}{\cosh(t+t_0)} \sinh(t)\right), \end{aligned} \tag{4.1}$$

then we get the speed υ_α and the tangent vector field T_α of α as

$$\upsilon_\alpha = \frac{a}{\cosh^2(t+t_0)}, \quad T_\alpha = (-\sinh(t_0), 0, \cosh(t_0)), \tag{4.2}$$

and therefore, the curvature κ_α of α is zero. Consequently, α cannot be a spacelike rectifying curve, as the definition of rectifying curve requires that its curvature non-zero. Therefore, not all spacelike curves that are dilations of unit speed spacelike curve $y(t)$ on H_0^2 of the type $\alpha(t) = \frac{a}{\cosh(t+t_0)} y(t)$ are rectifying curves. Therefore, the following theorem can be given according to the above findings.

Theorem 4.1. Let $y(t)$ be a unit speed spacelike curve on H_0^2 centered at the origin $0 \in E_1^3$ and let $\alpha(t) = \frac{a}{\cosh(t+t_0)} y(t)$ be a dilation of α spacelike rectifying curve with a timelike position vector and a timelike rectifying plane. The Frenet-Serret apparatus of α:

$$T_\alpha = -\sinh(t+t_0) y + \cosh(t+t_0) y',$$

$$N_\alpha = y \times y',$$
$$B_\alpha = -\cosh(t+t_0)y + \sinh(t+t_0)y',$$
$$\kappa_\alpha = \frac{1}{a}\cosh^3(t+t_0)\sqrt{\kappa_y^2+1},$$
$$\tau_\alpha = \frac{1}{a}\cosh^2(t+t_0)\sinh(t+t_0)\sqrt{\kappa_y^2+1}$$

where κ_y is the curvature of the unit speed spacelike curve $y(t)$.

Proof. By according to hypothesis of the theorem 4.1, the speed of α is given by, $v_\alpha = \frac{a}{\cosh^2(t+t_0)}$). Since $\{y, y', y \times y'\}$ is an orthonormal frame of E_1^3 along $y(s)$, we get,

$$T_\alpha = -\sinh(t+t_0)y + \cosh(t+t_0)y'.$$

Let s be arc-length parameter for α; then we have,

$$\frac{ds}{dt} = \frac{a}{\cosh^2(t+t_0)}.$$

By differentiating of equation T_α and using equation (2.4) and Frenet-Serret formulae, we get

$$\kappa_\alpha\left(\frac{a}{\cosh^2(t+t_0)}\right)N_\alpha = \cosh(t+t_0)hy \times y',$$

with $h = g(y'', y \times y')$. Therefore, by according to equation (2.6), we get,

$$\kappa_\alpha = \frac{1}{a}\cosh^3(t+t_0)\sqrt{{\kappa_y}^2+1},$$

and

$$N_\alpha = y \times y'.$$

Now, using $B_\alpha = T_\alpha \times N_\alpha$ we get,

$$B_\alpha = -\cosh(t+t_0)y + \sinh(t+t_0)y',$$

with $y \times (y \times y') = -y'$ and $y' \times (y \times y') = -y$.
After differentiating the equation above and using equation (2.4) and Frenet-Serret formulae we get,

$$\tau_\alpha N_\alpha\left(\frac{a}{\cosh^2(t+t_0)}\right) = \sinh(t+t_0)hy \times y',$$

and it leads to

$$\tau_\alpha = \frac{1}{a}\cosh^2(t+t_0)\sinh(t+t_0)\sqrt{{\kappa_y}^2+1}.$$

Q.E.D.

Theorem 4.2. Let $y(t)$ be a unit speed timelike curve on S_1^2 centered at the origin $0 \in E_1^3$ and let $\alpha(t) = \dfrac{a}{\sinh(t+t_0)} y(t)$ be a dilation of α with a spacelike position vector and a timelike rectifying plane. The Frenet-Serret apparatus of α:

$$T_\alpha = -\cosh(t+t_0)y + \sinh(t+t_0)y',$$

$$N_\alpha = y \times y',$$

$$B_\alpha = -\sinh(t+t_0)y + \cosh(t+t_0)y',$$

$$\kappa_\alpha = \frac{1}{a}\sinh^3(t+t_0)\sqrt{\kappa_y^2+1},$$

$$\tau_\alpha = \frac{1}{a}\sinh^2(t+t_0)\cosh(t+t_0)\sqrt{\kappa_y^2+1}$$

where κ_y is the curvature of the unit speed timelike curve $y(t)$.

Proof. The proof is made in a similar way to theorem 4.1. Q.E.D.

Corollary 4.3. Let $y(t)$ be a unit speed spacelike (resp. timelike) curve on H_0^2(resp. S_1^2) that is not an arc of the great circle, then $\alpha(t) = \dfrac{a}{\cosh(t+t_0)} y(t)$ (resp. $\alpha(t) = \dfrac{a}{\sinh(t+t_0)} y(t)$) is a rectifying spacelike curve.

5 Centrodes of unit speed spacelike curves

Let $\alpha : I \to E_1^3$ be a unit speed spacelike curve with Frenet-Serret apparatus $\{\kappa_\alpha, \tau_\alpha, T_\alpha, N_\alpha, B_\alpha\}$ and let $D_\alpha : I \to E_1^3$ be the centrode of α defined by

$$D_\alpha = \tau_\alpha T_\alpha - \kappa_\alpha B_\alpha. \tag{5.1}$$

By differentiating the expression of D_α and using Frenet-Serret formulae we get

$$D_\alpha' = \tau_\alpha' T_\alpha - \kappa_\alpha' B_\alpha.$$

Therefore, the speed υ_D of centrode D_α is given by

$$\upsilon_D = \sqrt{|(\tau_\alpha')^2 - \varepsilon(\kappa_\alpha')^2|}. \tag{5.2}$$

Hence the unit tangent vector field T_D of the centrode is given by

$$T_D = \frac{\tau_\alpha'}{\upsilon_D} T_\alpha - \frac{\kappa_\alpha'}{\upsilon_D} B_\alpha. \tag{5.3}$$

Let s be the arc-length parameter and let κ_D denote the curvature of the centrode. Then, by differentiating equation (5.3), we get,

$$\upsilon_D \kappa_D N_D = \left(\frac{\tau_\alpha'}{\upsilon_D}\right)' T_\alpha + \left(\frac{\tau_\alpha' \kappa_\alpha - \kappa_\alpha' \tau_\alpha}{\upsilon_D}\right) N_\alpha - \left(\frac{\kappa_\alpha'}{\upsilon_D}\right)' B_\alpha. \tag{5.4}$$

Proposition 5.1. Let $\alpha : I \to E_1^3$ be a unit speed spacelike curve whose curvature κ_α and torsion τ_α satisfy $\kappa_\alpha \neq 0$ and $|(\tau_\alpha^{'})^2 - \varepsilon(\kappa_\alpha^{'})^2| \neq 0$. Then α is a helix if and only if its centrode is a line segment.

Proof. If the centrode α is a line segment, then equation (5.4) gives $\tau_\alpha^{'}\kappa_\alpha = \kappa_\alpha^{'}\tau_\alpha$ and this shows that $\frac{\tau_\alpha}{\kappa_\alpha}$ is a constant. Hence α is a helix.
Conversely, if α is a helix, then we have $\tau_\alpha = c\kappa_\alpha$ for some constant $c \neq 0$, and thus equation (5.4) gives $\kappa_D = 0$. Hence α is a line segment. Q.E.D.

Theorem 5.2. We have:
(a) Let $\alpha = \alpha(t)$ be a unit speed spacelike curve in E_1^3 whose curvature κ_α and torsion τ_α satisfy $\kappa_\alpha, \tau_\alpha \neq 0$ and $|(\tau_\alpha^{'})^2 - \varepsilon(\kappa_\alpha^{'})^2| \neq 0$. Suppose that α is not a helix. Then the centrode $D_\alpha = \tau_\alpha T_\alpha - \kappa_\alpha B_\alpha$ of α is a rectifying spacelike curve if and only if κ_α and τ_α satisfy a non-homogeneus linear equation $a\kappa_\alpha - b\tau_\alpha = c$, so that a, b, c are constants with $a^2 + b^2 \neq 0$ and $c \neq 0$.
(b) If $\kappa_\alpha^{'} \neq 0$ for $\kappa_\alpha^{'} \neq \mp\tau_\alpha^{'}$ and if the centrode $D_\alpha(t)$ of $\alpha(t)$ is a spacelike rectifying curve, then the Frenet-Serret apparatus $\{\kappa_D, \tau_D, T_D, N_D, B_D\}$ of the centrode satisfies,

$$T_D = \frac{\widehat{c}}{\sqrt{|\widehat{c}^2 - \varepsilon|}} T_\alpha - \frac{1}{\sqrt{|\widehat{c}^2 - \varepsilon|}} B_\alpha,$$
$$N_D = N_\alpha,$$
$$B_D = \frac{1}{\sqrt{|\widehat{c}^2 - \varepsilon|}} T_\alpha + \frac{\widehat{c}}{\sqrt{|\widehat{c}^2 - \varepsilon|}} B_\alpha,$$
$$\kappa_D = \frac{(\widehat{c}\kappa_\alpha - \tau_\alpha)}{\kappa_\alpha^{'}(|\widehat{c}^2 - \varepsilon|)},$$
$$\tau_D = \frac{-\widehat{c}\tau_\alpha + \kappa_\alpha}{\kappa_\alpha^{'}(|\widehat{c}^2 - \varepsilon|)},$$

where $\widehat{c} = \frac{a}{b}$ for $\widehat{c} \neq \mp 1$ and a, b are defined as in statement (a).
(c) If $\tau_\alpha^{'} \neq 0$ for $\kappa_\alpha^{'} \neq \mp\tau_\alpha^{'}$ and if the centrode $D(t)$ is a rectifying spacelike curve, then the Frenet-Serret apparatus $\{\kappa_D, \tau_D, T_D, N_D, B_D\}$ of the centrode satisfies,

$$T_D = \frac{1}{\sqrt{|1 + \overline{c}^2\varepsilon|}} T_\alpha + \frac{\overline{c}}{\sqrt{|1 - \overline{c}^2\varepsilon|}} B_\alpha,$$
$$N_D = N_\alpha,$$
$$B_D = \frac{\overline{c}}{\sqrt{|1 - \overline{c}^2\varepsilon|}} T_\alpha + \frac{1}{\sqrt{|1 - \overline{c}^2\varepsilon|}} B_\alpha,$$
$$\kappa_D = \frac{-\overline{c}\tau_\alpha + \kappa_\alpha}{\tau_\alpha^{'}|1 - \overline{c}^2\varepsilon|},$$
$$\tau_D = \frac{\overline{c}\kappa_\alpha - \tau_\alpha}{\tau_\alpha^{'}|1 - \overline{c}^2\varepsilon|},$$

where $\overline{c} = \frac{b}{a}$ for $\overline{c} \neq \mp 1$ and a, b are defined as in statement (a).

Proof. Let $\alpha(t)$ be a unit speed spacelike curve in E_1^3 whose curvature κ_α and torsion τ_α satisfy $\kappa_\alpha, \tau_\alpha \neq 0$ and $|(\tau_\alpha^{'})^2 - \varepsilon(\kappa_\alpha^{'})^2| \neq 0$. Suppose that α is not a helix. By according to equations (5.2) and (5.4), we get,

$$\upsilon_D \kappa_D \langle N_D, D_\alpha \rangle = \tau_\alpha (\frac{\tau_\alpha^{'}}{\upsilon_D})' - \varepsilon \kappa_\alpha (\frac{\kappa_\alpha^{'}}{\upsilon_D})'. \tag{5.5}$$

Since α is not a helix, $\frac{\tau_\alpha}{\kappa_\alpha}$ is non-constant. Hence, by according to proposition 5.1, we have $\kappa_D \neq 0$ and as $\sqrt{|(\tau_\alpha^{'})^2 - \varepsilon(\kappa_\alpha^{'})^2|} \neq 0$. Therefore equation (5.5) implies that,

$$\tau_\alpha (\frac{\tau_\alpha^{'}}{\sqrt{|(\tau_\alpha^{'})^2 - \varepsilon(\kappa_\alpha^{'})^2|}})' - \varepsilon \kappa_\alpha (\frac{\kappa_\alpha^{'}}{\sqrt{|(\tau_\alpha^{'})^2 - \varepsilon(\kappa_\alpha^{'})^2|}})' = 0, \tag{5.6}$$

holds identically if and only if $\langle N_D, D_\alpha \rangle = 0$ holds identically. Consequently, the centrode D_α of α is spacelike rectifying curve if and only if equation (5.6) holds.
Now we shall solve differential equation (5.6). Since $|(\tau_\alpha^{'})^2 - \varepsilon(\kappa_\alpha^{'})^2| \neq 0$, we have $(\tau_\alpha^{'})^2 - \varepsilon(\kappa_\alpha^{'})^2 \neq 0$. Consequently, we have Case 1,2,3 and 4.
Case (1): Let us assume that $\varepsilon = -1$ and $\kappa_\alpha^{'} \neq 0$. In this case we define a function g_1 by

$$g_1(t) = \tan^{-1}\left(\frac{\tau_\alpha^{'}}{\kappa_\alpha^{'}}\right). \tag{5.7}$$

From (5.7), we get

$$\sin g_1(t) = \frac{\tau_\alpha^{'}}{\sqrt{(\tau_\alpha^{'})^2 + (\kappa_\alpha^{'})^2}}, \cos g_1(t) = \frac{\kappa_\alpha^{'}}{\sqrt{(\tau_\alpha^{'})^2 + (\kappa_\alpha^{'})^2}}. \tag{5.8}$$

Hence, the equation (5.6) is equivalent to

$$(\tau_\alpha(\cos g_1(t)) - \kappa_\alpha(\sin g_1(t)))g_1^{'}(t) = 0. \tag{5.9}$$

If $\tau_\alpha(\cos g_1(t)) - \kappa_\alpha(\sin g_1(t)) = 0$ we have,

$$\frac{\tau_\alpha}{\kappa_\alpha} = \tan g_1(t) = \frac{\tau_\alpha^{'}}{\kappa_\alpha^{'}},$$

which implies that $\frac{\tau_\alpha}{\kappa_\alpha}$ is a constant. However, this is impossible since the spacelike curve α is not a helix. Therefore, we obtain $g_1^{'}(t) = 0$ from (5.9), and thus $g_1(t)$ is a constant. Consequently, $\tau_\alpha^{'} = c_1 \kappa_\alpha^{'}$ for some constant c_1. If we put $c_1 = \frac{a}{b}$ for constants a, b. Then, we obtain $a\kappa_\alpha - b\tau_\alpha = c$ for some constant c. Since $\frac{\tau_\alpha}{\kappa_\alpha}$ is non-constant, we must have $c \neq 0$ and hence $a^2 + b^2 \neq 0$.
Case (2): Let us assume that $\varepsilon = -1$ and $\tau_\alpha^{'} \neq 0$. In this case we define a function g_2 by

$$g_2(t) = \tan^{-1}\left(\frac{\kappa_\alpha^{'}}{\tau_\alpha^{'}}\right). \tag{5.10}$$

From (5.10), we get

$$\sin g_2(t) = \frac{\kappa_\alpha^{'}}{\sqrt{(\tau_\alpha^{'})^2 + (\kappa_\alpha^{'})^2}}, \quad \cos g_2(t) = \frac{\tau_\alpha^{'}}{\sqrt{(\tau_\alpha^{'})^2 + (\kappa_\alpha^{'})^2}}. \tag{5.11}$$

Similarly, we know that equation (5.6) is equivalent to,

$$(-\tau_\alpha(\sin g_2(t)) + \kappa_\alpha(\cos g_2(t)))g_2^{'}(t) = 0.$$

If $-\tau_\alpha(\sin g_2(t)) + \kappa_\alpha(\cos g_2(t) = 0$ we have

$$\frac{\kappa_\alpha}{\tau_\alpha} = \tan g_2(t) = \frac{\kappa_\alpha^{'}}{\tau_\alpha^{'}}.$$

Now, by appyling a similar argument as Case 1, we obtain $a\kappa_\alpha - b\tau_\alpha = c$ for some constants a, b, c with $a^2 + b^2 \neq 0$ and $c \neq 0$.

Case (3): Let us assume that $\varepsilon = 1$ and $\tau_\alpha^{'} \neq \pm\kappa_\alpha^{'} \neq 0$. In this case we define a function g_3 by

$$g_3(t) = \tanh^{-1}\left(\frac{\tau_\alpha^{'}}{\kappa_\alpha^{'}}\right) \tag{5.12}$$

From (5.12), we get

$$\sinh g_3(t) = \frac{\tau_\alpha^{'}}{\sqrt{|(\tau_\alpha^{'})^2 - (\kappa_\alpha^{'})^2|}}, \quad \cosh g_3(t) = \frac{\kappa_\alpha^{'}}{\sqrt{|(\tau_\alpha^{'})^2 - (\kappa_\alpha^{'})^2|}}. \tag{5.13}$$

Similarly, we know that equation (5.6) is equivalent to,

$$(\tau_\alpha \cosh g_3(t) - \kappa_\alpha \sinh g_3(t))g_3^{'}(t) = 0.$$

If $\tau_\alpha(\cosh g_3(t)) - \kappa_\alpha(\sinh g_3(t)) = 0$ we have

$$\frac{\tau_\alpha}{\kappa_\alpha} = \tanh g_3(t) = \frac{\tau_\alpha^{'}}{\kappa_\alpha^{'}}.$$

Now, by appyling a similar argument as Case 1,2 we obtain $a\kappa_\alpha - b\tau_\alpha = c$ for some constants a, b, c with $a^2 + b^2 \neq 0$ and $c \neq 0$.

Case (4): Let us assume that $\varepsilon = 1$ and $\tau_\alpha^{'} \neq \pm\kappa_\alpha^{'} \neq 0$. In this case we define a function g_4 by

$$g_4(t) = \tanh^{-1}\left(\frac{\kappa_\alpha^{'}}{\tau_\alpha^{'}}\right). \tag{5.14}$$

From (5.14), we get

$$\sinh g_4(t) = \frac{\kappa_\alpha^{'}}{\sqrt{|(\tau_\alpha^{'})^2 - (\kappa_\alpha^{'})^2|}}, \quad \cosh g_4(t) = \frac{\tau_\alpha^{'}}{\sqrt{|(\tau_\alpha^{'})^2(\kappa_\alpha^{'})^2|}}. \tag{5.15}$$

Similarly, we know that equation (5.6) is equivalent to,

$$(\tau_\alpha \sinh g_4(t) - \kappa_\alpha \cosh g_4(t))g_4'(t) = 0.$$

If $\tau_\alpha(\sinh g_4(t)) - \kappa_\alpha(\cosh g_4(t)) = 0$ we have

$$\frac{\kappa_\alpha}{\tau_\alpha} = \tanh g_4(t) = \frac{\kappa_\alpha'}{\tau_\alpha'}.$$

Now, by appyling a similar argument as Case 1,2,3, we obtain $a\kappa_\alpha - b\tau_\alpha = c$ for some constants a, b, c with $a^2 + b^2 \neq 0$ and $c \neq 0$.

Conversely, if κ_α and τ_α satisfy $a\kappa_\alpha - b\tau_\alpha = c$ for some constants a, b, c satisfying $a^2 + b^2 \neq 0$ and $c \neq 0$, then κ_α and τ_α satisfy the differential equation (5.6). This proves statement(a).

Assume that the centrode $D_\alpha(t)$ is a spacelike rectifying curve. Then statement (a) implies that the curvature and torsion of $\alpha(t)$ satisfy a non-homogeneous linear equation

$$a\kappa_\alpha - b\tau_\alpha = c,$$

for some constants a, b, c satisfying $a^2 + b^2 \neq 0$ and $c \neq 0$. In particular, we have (i) $b \neq 0$ or (ii) $a \neq 0$.

Case (i): $b \neq 0$. We find from (5.12) that

$$\tau_\alpha' = \widehat{c}\kappa_\alpha', \quad \widehat{c} = \frac{a}{b} \neq \pm 1. \tag{5.16}$$

Therefore we get from (5.2) and (5.3)

$$\upsilon_D = \kappa_\alpha' \sqrt{|\widehat{c}^2 - \varepsilon|}, \tag{5.17}$$

$$T_D = \frac{\widehat{c}}{\sqrt{|\widehat{c}^2 - \varepsilon|}} T_\alpha - \frac{1}{\sqrt{|\widehat{c}^2 - \varepsilon|}} B_\alpha.$$

Consequently, equation (5.5) reduces to

$$\kappa_\alpha' \sqrt{|\widehat{c}^2 - \varepsilon|}\kappa_D N_D = \frac{\widehat{c}\kappa_\alpha - \tau_\alpha}{\sqrt{|\widehat{c}^2 - \varepsilon|}} N_\alpha,$$

$$\kappa_D = \frac{(\widehat{c}\kappa_\alpha - \tau_\alpha)}{\kappa_\alpha'(|\widehat{c}^2 - \varepsilon|)}, \quad N_D = N_\alpha. \tag{5.18}$$

By using equations (5.15), (5.16) and $B_D = T_D \times N_D$ we get,

$$B_D = -\frac{1}{\sqrt{|\widehat{c}^2 - \varepsilon|}} T_\alpha + \frac{\widehat{c}}{\sqrt{|\widehat{c}^2 - \varepsilon|}} B_\alpha,$$

which on differentiation and using Frenet-Serret formulae gives,

$$\tau_D = \frac{-\widehat{c}\tau_\alpha + \kappa_\alpha}{\kappa_\alpha'(|\widehat{c}^2 - \varepsilon|)}.$$

This proves statement (b).

Case (ii): $a \neq 0$. From (5.14) we have $\kappa_\alpha' = \bar{c}\tau_\alpha', \bar{c} = \frac{b}{a} \neq \pm 1$. Thus we may apply a method similar to Case (i) to obtain statement (c). Q.E.D.

Remark 5.3. The condition $a\kappa_\alpha - b\tau_\alpha = c$ with $a^2 + b^2 \neq 0$ and $c \neq 0$ given in theorem 4.1(a) has been used by Lucas and Ortega-Yages in [13] for their study of Bertrand curves in the Euclidean 3-space E^3 or Lorentz-Minkowski 3-space L^3.

Remark 5.4. Let $\alpha(t)$ be the unit speed spacelike curve of theorem 5.1 with $\kappa_{\alpha}' \neq 0 \quad (\tau_\alpha' \mp \neq \kappa_\alpha')$ such that the centrode D_α of α is a spacelike rectifying curve. Then as $\tau_\alpha' = \widehat{c}\kappa_\alpha'$ with $\widehat{c}\mp \neq 1$, we get $\widehat{c}\kappa_\alpha = \tau_\alpha + c_1$ for a constant $c_1 \neq 0$ (as $\frac{\tau_\alpha}{\kappa_\alpha}$ is non-constant), that is, $\widehat{c}\kappa_\alpha - \tau_\alpha = c_1$. Moreover, the arc-length parameter s of the spacelike rectifying curve D_α satisfies $\frac{ds}{dt} = \kappa_\alpha' \sqrt{|\widehat{c}^2 - \varepsilon|}$ which gives

$$s = \kappa_\alpha \sqrt{|\widehat{c}^2 - \varepsilon|} + b,$$

for a constant b. Thus after using the expressions of curvature and torsion of D_α we obtain,

$$\frac{\tau_D}{\kappa_D} = \frac{-\widehat{c}\tau_\alpha + \kappa_\alpha}{\widehat{c}\kappa_\alpha - \tau_\alpha} = \frac{-\widehat{c}\tau_\alpha + \kappa_\alpha}{c_1} = \frac{-\widehat{c}}{c_1}(\widehat{c}\kappa - c_1) + \frac{\kappa_\alpha}{c_1},$$

$$= (\frac{-\widehat{c}^2 + 1}{c_1})\kappa_\alpha + \widehat{c} = \frac{-\widehat{c}^2 + 1}{c_1}(\frac{s - b}{\sqrt{|\widehat{c}^2 - \varepsilon|}}) + \widehat{c},$$

$$= As + B,$$

where $A \neq 0, \quad B$ are constants. Thus, the ratio $\frac{\tau_D}{\kappa_D}$ is a linear function of the arc-length s as required by a spacelike rectifying curve (cf. [2, Theorem 2]).
Also, we get

$$|\tau_\alpha^2 - \varepsilon\kappa_\alpha^2| = |(\widehat{c}\kappa_\alpha - c_1)^2 - \varepsilon\kappa\alpha^2| = |\kappa_\alpha^2(\widehat{c}^2 - \varepsilon) - 2\widehat{c}c_1\kappa_\alpha + c_1^2|,$$

$$= |\widehat{c}^2 - \varepsilon|\frac{(s-b)^2}{|\widehat{c}^2 - \varepsilon|} - 2c_1\widehat{c}\frac{(s-b)}{\sqrt{|\widehat{c}^2 - \varepsilon|}} + c_1^2,$$

$$= s^2 + \lambda_1 s + \lambda_2,$$

where λ_1, λ_2 are constants in E_1^3. Therefore the distance function $f(s) = \|D_\alpha\|$ of the spacelike rectifying curve D_α satisfies $f(s) = \sqrt{s^2 + \lambda_1 s + \lambda_2}$, as required by a spacelike rectifying curve.

Note that similar arguments hold for a unit speed spacelike curve $\alpha(t)$ with $\tau_\alpha' \neq 0$ (instead of $\kappa_\alpha' \neq 0$) and with a spacelike rectifying centrode.

Corollary 5.5. Let $\alpha : I \to E_1^3$ be a unit speed spacelike curve whose curvature κ_α and torsion τ_α satisfy $\kappa_\alpha^{'} \neq 0$, $\frac{\tau_\alpha}{\kappa_\alpha}$ non-constant, and $\tau_\alpha^{'} = \widehat{c}\kappa_\alpha^{'}, \widehat{c} \neq \mp 1$ being a constant. Then the centrode D_α of α is a spacelike rectifying curve with curvature $\kappa_D = \pm\frac{a}{\kappa_\alpha^{'}}$ and torsion $\tau_D = \frac{b_1 s + c_1}{\kappa_\alpha^{'}}$ for some constants $a, b_1 \neq 0$ and c_1.

Proof. Under the hypothesis of the theorem, we have

$$\frac{\tau_D}{\kappa_D} = \frac{-\widehat{c}\tau_\alpha + \kappa_\alpha}{\widehat{c}\kappa_\alpha - \tau_\alpha},$$

which implies

$$1 - (\frac{\tau_D}{\kappa_D})^2 = \frac{\widehat{c}^2 - 1}{(\widehat{c}\kappa_\alpha - \tau_\alpha)^2}(\kappa_\alpha^2 - \tau_\alpha^2). \tag{5.19}$$

Now, as $\tau_\alpha^{'} = \widehat{c}\kappa_\alpha^{'}$, we have $\widehat{c}\kappa_\alpha - \tau_\alpha = c_0$, where c_0 is non-zero constant (as $\frac{\tau_\alpha}{\kappa_\alpha}$ is non-constant), and from equation (5.19), we get

$$\frac{\kappa_D^2 - \tau_D^2}{\kappa_\alpha^2 - \tau_\alpha^2} = \frac{\widehat{c}^2 - 1}{c_0^2}\kappa_D^2. \tag{5.20}$$

By according to equation (5.19), we find

$$\kappa_D^2 - \tau_D^2 = \frac{\widehat{c}^2 - 1}{(\kappa_\alpha^{'})^2(\widehat{c}^2 - \varepsilon)^2}(\kappa_\alpha^2 - \tau_\alpha^2),$$

that is,

$$\frac{\kappa_D^2 - \tau_D^2}{\kappa_\alpha^2 - \tau_\alpha^2} = \frac{\widehat{c}^2}{(\kappa_\alpha^{'})^2(\widehat{c}^2 - \varepsilon)}.$$

After combining this equation with equation (5.20), we get

(i) If T_α, N_α are spacelike vectors and B_α is a timelike vector, i.e. $\varepsilon = 1$, then we have

$$\kappa_D^2 = \frac{c_0^2}{(\kappa_\alpha^{'})^2(\widehat{c}^2 - 1)^2} = \frac{c_2^2}{(\kappa_\alpha^{'})^2},$$

where c_2 is a non-zero constant. Therefore we obtain the required expression for the curvature κ_D of the spacelike rectifying curve D_α.

(ii) If T_α, B_α are spacelike vectors and N_α is a timelike vector, i.e. $\varepsilon = -1$, then we have

$$\kappa_D^2 = \frac{c_0^2}{(\kappa_\alpha^{'})^2(\widehat{c}^2 + 1)^2} = \frac{c_3^2}{(\kappa_\alpha^{'})^2},$$

where c_3 is a non-zero constant. Therefore we obtain the required expression for the curvature κ_D of the spacelike rectifying curve D_α.

The arc-length function s of the spacelike rectifying curve is given by $s = \kappa_\alpha\sqrt{|\widehat{c}^2 - \varepsilon|} + c_3$ for a constant c_3 and consequently, using the expression for torsion τ_D in statement (b) of theorem 5.1, we derive that

$$\tau_D = \frac{-\widehat{c}\tau_\alpha + \kappa_\alpha}{\kappa_\alpha^{'}|\widehat{c}^2 - \varepsilon|} = \frac{-\widehat{c}(\widehat{c}\kappa_\alpha - c_0) + \kappa_\alpha}{\kappa_\alpha^{'}|\widehat{c}^2 - \varepsilon|},$$

$$\frac{(-\widehat{c}^2 + 1)\kappa_\alpha + \widehat{c}c_0}{\kappa_\alpha^{'}|\widehat{c}^2 - \varepsilon|} = \frac{1}{\kappa_\alpha^{'}}(-\widehat{c}^2 + 1)[\frac{(s - c_3)}{\sqrt{|\widehat{c}^2 - \varepsilon|}|\widehat{c}^2 - \varepsilon|} + \frac{\widehat{c}c_0}{\widehat{c}^2 - \varepsilon}].$$

(i) If T_α, N_α are spacelike vectors and B_α is a timelike vector, i.e. $\varepsilon = 1$, then we have

$$\tau_D = -\frac{1}{\kappa_\alpha^{'}}[\frac{(s - c_3)}{\sqrt{\widehat{c}^2 - 1}} - \widehat{c}c_0,$$

$$= \frac{c_4 s + c_5}{\kappa_\alpha^{'}},$$

where $c_4 \neq 0$ and c_5 are constants.
(ii) If T_α, B_α are spacelike vectors and N_α is a timelike vector, i.e. $\varepsilon = -1$, then we have

$$\frac{1}{\kappa_\alpha^{'}}[\frac{(-\widehat{c}^2 + 1)(s - c_3)}{\sqrt{\widehat{c}^2 + 1}(\widehat{c}^2 + 1)} - \frac{\widehat{c}c_0}{(\widehat{c}^2 + 1)},$$

$$= \frac{c_6 s + c_7}{\kappa_\alpha^{'}},$$

where $c_6 \neq 0$ and c_7 are constants. Q.E.D.

Remark 5.6. A result similar to corollary 5.1 holds for a unit speed spacelike curve satisfying $\tau_\alpha^{'} \neq 0$, $\frac{\tau_\alpha}{\kappa_\alpha}$ non-constant, and $\kappa_\alpha^{'} = \overline{c}\tau_\alpha^{'}$ with constant $\overline{c}$.

6 Dilated centrodes as spacelike rectifying curves

Finally, we study the dilated centrode of a unit speed twisted spacelike curve $\alpha : I \to E_1^3$ with $\kappa_\alpha > 0, \tau_\alpha \neq 0$. The dilated centrode of $\alpha(t)$ is defined by

$$\beta(t) = \rho_\alpha(t)D_\alpha(t) = \frac{\tau_\alpha(t)}{\kappa_\alpha(t)}T_\alpha - B_\alpha, \tag{6.1}$$

where $D_\alpha = \tau_\alpha T_\alpha - \kappa_\alpha B_\alpha$ is the centrode and $\rho_\alpha = \kappa_\alpha^{-1}$ is the radius of curvature of α spacelike curve.

Theorem 6.1. Let $\alpha : I \to E_1^3$ be a unit speed spacelike curve with curvature $\kappa_\alpha > 0$. If α is neither a planar spacelike curve nor a helix, then the dilated centrode $\beta(t) = \rho_\alpha(t)D_\alpha(t)$ of is α spacelike rectifying curve.

Proof. Let $\alpha : I \to E_1^3$ be a unit speed spacelike with curvature $\kappa_\alpha > 0$. Suppose that α is neither a planar spacelike curve nor a helix. Consider the dilated centrode of defined by (6.1). By differentiating equation (6.1) and using Frenet-Serret formulae, we get

$$(\beta)^{'} = (\rho_\alpha\tau_\alpha)^{'}T_\alpha,$$

and since α is not a helix, we have $(\rho_\alpha \tau_\alpha)' \neq 0$. Thus the dilated centrode $\beta(t)$ is a regular spacelike curve whose speed υ_β and unit tangent vector field T_β are given respectively by

$$\begin{aligned} \upsilon_\beta &= (\rho_\alpha \tau_\alpha)', \\ T_\beta &= T_\alpha. \end{aligned} \tag{6.2}$$

Let $\{\kappa_\beta, \tau_\beta, T_\beta, N_\beta, B_\beta\}$ be the Frenet-Serret apparatus of β. Then after differentiating equation (6.2) we find,

$$\kappa_\beta (\rho_\alpha \tau_\alpha)' N_\beta = \kappa_\alpha N_\alpha,$$

and

$$\kappa_\beta = \frac{\kappa_\alpha}{(\rho_\alpha \tau_\alpha)'}, \quad N_\beta = N_\alpha. \tag{6.3}$$

Using equations (6.1) and (6.3) we find that $\langle \beta, N_\beta \rangle = 0$, that is, the dilated centrode β is a spacelike rectifying curve. This proves the theorem.
Note that if follows from (6.2) and (6.3) that $B_\beta = B_\alpha$ and that,

$$\tau_\beta = \frac{\tau_\alpha}{(\rho_\alpha \tau_\alpha)'}. \tag{6.4}$$

Q.E.D.

Remark 6.2. Let $\alpha : I \to E_1^3$ be a unit speed spacelike curve with Frenet-Serret apparatus $\{\kappa_\alpha, \tau_\alpha, T_\alpha, N_\alpha, B_\alpha\}$, $\kappa_\alpha > 0, \tau_\alpha \neq 0$ and $(\rho_\alpha \tau_\alpha)' \neq 0$. Then according to theorem 6.1, the spacelike curve $\beta(t) = \rho_\alpha \tau_\alpha T_\alpha - B_\alpha$ is a spacelike rectifying curve whose Frenet-Serret apparatus $\{\kappa_\beta, \tau_\beta, T_\beta, N_\beta, B_\beta\}$ is given by equations (6.2) and (6.3). If s is the arc-length of β, then by equation (6.2) we have $s = \pm \rho_\alpha \tau_\alpha + c$ for a constant c. The distance function $f(s) = \|\beta(s)\|$ in view of equation (6.1) is given by

$$f(s) = \sqrt{|(\rho_\alpha \tau_\alpha)^2 - \varepsilon|} = \sqrt{(s-c)^2 + (-\varepsilon)} = \sqrt{s^2 + c_1 s + c_2},$$

where $c_1 = -2c, c_2 = c^2 - \varepsilon$. Hence the distance function $f(s)$ of the spacelike rectifying curve β has the form described in [2].
Similarly, using equation (6.3) and (6.4), we get,

$$\frac{\tau_\beta}{\kappa_\beta} = \frac{\tau_\alpha}{\kappa_\alpha} = \mp(s-c) = as + b,$$

where $a \neq 0, b$ are constants. Therefore the ratio of torsion and curvature of the spacelike rectifying curve β is the linear function of the arc-length [2].

Example 6.3. Let the curve $\alpha : I \to E_1^3$ defined by

$$\alpha(t) = (\cosh t, \sinh t, \frac{t^2}{2}).$$

The speed, unit tangent vector field, and curvature of α given respectively by

$$\upsilon_\alpha = \sqrt{t^2+1}, \quad T_\alpha = \frac{1}{\sqrt{t^2+1}}(\sinh(t), \cosh t, t), \quad \kappa_\alpha = \frac{\sqrt{t^2+2}}{(t^2+1)^{\frac{3}{2}}}$$

for $t > 1$ and the principal normal vector field

$$N_\alpha = \frac{1}{\sqrt{(t^2+1)(t^2+2)}}((t^2+1)\cosh t - t\sinh t, (t^2+1)\sinh t - t\cosh t, 1).$$

The binormal vector field and torsion of α are

$$B_\alpha = \frac{1}{\sqrt{t^2+2}}(t\sinh t - \cosh t, t\cosh t - \sinh t, -1) \quad , \tau_\alpha = \frac{-t}{t^2+2}.$$

Consequently, we have $\kappa_\alpha > 0$ and that α is neither a planar curve nor a helix. Hence, by theorem 6.1, the dilated centrode

$$\beta(t) = \frac{1}{\kappa_\alpha} D_\alpha(t) = \left(\frac{\tau_\alpha}{\kappa_\alpha}\right) T_\alpha - B_\alpha,$$

of α is a spacelike rectifying curve.

References

[1] P. Appell, Traite de Mecanique Rationnelle vol. 1. 6th ed. Paris, France: Gauthier-Villars, (1941).

[2] B. Y. Chen, *When does the position vector of a space curve always lie in its rectifying plane?* Amer Math Monthly (2003); 110: 147-152.

[3] B. Y. Chen, *Rectifying curves and geodesics on a cone in the Euclidean 3-space.* Tamkang J Math (2017); 48: 209-214.

[4] B. Y. Chen, F. Dillen, *Rectifying curves as centrodes and extremal curves.* Bull Inst Math Acad Sinica (2005); 33: 77-90.

[5] S. Deshmukh, B. Y. Chen, S. H. Alshammari, *On rectifying curves in Euclidean 3-space.* Turk J Math (2018);42(2): 609-620.

[6] S. D. Gertzbein, J. Seligman, R. Holtby, K. W. Chan, N. Ogston, A. Kapasouri, M. Tile, B. Cruickshank, *Centrode patterns and segmental instability in degenerative disk disease.* Spine (1985); 10: 257-261.

[7] K. Ilarslan, E. Nesovic, *Some characterizations of rectifying curves in the Euclidean space E^4.* Turk J Math (2008); 32: 21-30.

[8] K. Ilarslan, E. Nesovic, *On rectifying curves as centrodes and extremal curves in the Minkowski 3-space.* Novi Sad J. Math (2007),37(1): 53-64.

[9] K. Ilarslan, E. Nesovic, *Petrovic-Torgasev. Some characterizations of rectifying curves in the Minkowski 3-space.* Novi Sad J. Math (2003); 33(2): 23-32.

[10] S. Izumiya, N. Takeuchi, *New special curves and developable surfaces.* Turk J Math (2004); 28:153-163.

[11] D. S. Kim, H. S. Chung, K. H. Cho, *Space curves satisfying* $\tau/\kappa = as + b$. Honam Math J (1993); 15: 5-9.

[12] W. Kuhnel, *Differential geometry:curves-surfaces-manifolds.* Weisbaden: Braunschweig (1999).

[13] P. Lucas, J. A. Ortega- Yagues, *A variational characterization and geometric integration for Bertrand curves.* J Math Phys (2013); 54: 043508, 12 pp.

[14] R. S. Millman, G. D. Parker, *Elements of Differential Geometry.* Englewood Cliffs, NJ, USA: Prentice-Hall, 1977.

[15] N. G. Ogston, G. J. King, S. D. Gertzbein, M. Tile, *Kapasouri A, Rubenstein J.D. Centrode patterns in the lumbar spine-base-line studies in normal subjects.* Spine (1986); 11: 591-595.

[16] J. Oprea *Differential Geometry and Its Applications*, 2nd edition. Washington, DC, USA: The Mathematical Association of America, (2007).

[17] P. J. Weiler, R. E. Bogoch. *Kinematics of the distal radioulnar joint in rheumatoid-arthritis-an in-vivo study using centrode analysis.* J Hand Surgery (1995); 20A: 937-943.

[18] H. Yeh, J. I. Abrams, *Principles of Mechanics of Solids and Fluids*, vol. 1. New York, NY, USA: McGraw-Hall, (1960).

[19] B. Yılmaz, I. Gök, Y. Yaylı, *Extended Rectifying Curves in Minkowski 3-Space.* Adv. Appl. Clifford Algebras 26 (2016); 861872.

DOI 10.2478/9788395793882-003

Framed normal curves in Euclidean space

Bahar Doğan Yazici[1], Sıddıka Özkaldı Karakuş[2] and Murat Tosun[3]

[1,2]Department of Mathematics, Bilecik Seyh Edebali University, 11230 Bilecik, Turkey

[3]Department of Mathematics, Sakarya University, 54100 Sakarya, Turkey

[1] Corresponding Author

E-mail: bahar.dogan@bilecik.edu.tr[1], siddika.karakus@bilecik.edu.tr[2], tosun@sakarya.edu.tr[3]

Abstract

A normal curve is a space curves which the position vector always lies in their normal plane. In this paper, we define framed normal curves and Frenet-type framed normal curves in Euclidean space. A framed base curve is a smooth curve with a moving frame which may have singular point [3]. We investigate the characterization of framed normal curves and reveal the relationships between framed normal curves and framed spherical curves.

2010 Mathematics Subject Classification. **53A04** 58K05
Keywords. normal curves, singular point, framed curves, framed spherical curves.

1 Introduction

The curves $\gamma : I \to \mathbb{R}^3$ for which the position vector γ always lie in their rectifying plane, are for simplicity called rectifying curves [1]. Similarly, the curves for which the position vector γ always lie in their osculating plane, are for simplicity called osculating curves; and finally, the curves for which the position vector always lie in their normal plane, are for simplicity called normal curves. K.Ilarslan and E. Nesovic investigated the notion of normal curves in Euclidean space and Minkowski space [5,6].

Many studies have been done using Frenet frames for regular curves in classical differential geometry. If space curves have singular points, the Frenet frames of these curves cannot be constructed. However, S.Honda and M.Takahashi gave the definition of framed curves [3]. A framed curve is a smooth curve with a moving frame which may have singular point. Moreover, T.Fukunaga and M. Takahashi gave existence conditions of framed curves for smooth curves in [2]. In addition, Y.Wang, D.Pei and R.Gao defined a adapted frame for framed curves and studied framed rectifying curves in Euclidean space [8].

Inspired by the above work, in this paper, we define framed normal curves and Frenet-type framed normal curves in Euclidean space. Also, the necessary and sufficient conditions are given for a framed curve to be a framed normal curves.

2 Framed curves in Euclidean space

A framed curve in the 3-dimensional Euclidean space is a smooth space curve with a moving frame which may have singular points, in detail see [3]. Let $\gamma : I \to \mathbb{R}^3$ be a curve with singular points. The set

$$\Delta_2 = \{\mu = (\mu_1, \mu_2) \in \mathbb{R}^3 \times \mathbb{R}^3 \mid \langle \mu_i, \mu_j \rangle = \delta_{ij}, \quad i, j = 1, 2\}$$

is an 3-dimensional smooth manifold. Suppose that $\mu = (\mu_1, \mu_2) \in \Delta_2$. A unit vector is defined by $\upsilon = \mu_1 \times \mu_2$.

Tbilisi Mathematical Journal Special Issue (IECMSA-2019), pp. 27–37.
Tbilisi Centre for Mathematical Sciences.

Received by the editors: 19 October 2019.
Accepted for publication: 03 January 2020.

Definition 2.1. We say that $(\gamma,\mu): I \to \mathbb{R}^3 \times \Delta_2$ is a framed curve is $\langle \gamma'(s), \mu_i(s)\rangle = 0$ for all $s \in I$ and $i = 1,2$. We also say that $\gamma : I \to \mathbb{R}^3$ is a framed base curve if there exists $\mu : I \to \Delta_2$ such that (γ,μ) is a framed curve [3].

In [2], the theorems of the existence and uniqueness for framed curves were shown as follows:

Theorem 2.2 (The Existence Theorem). Let $(l,m,n,\alpha) : I \to \mathbb{R}^4$ be a smooth mapping. There exists a framed curve $(\gamma,\mu) : I \to \mathbb{R}^3 \times \Delta_2$ whose associated curvature is (l,m,n,α).

Theorem 2.3 (The Uniqueness Theorem). Let (γ,μ), $(\overline{\gamma},\overline{\mu})$ be framed curves whose curvatures (l,m,n,α) and $(\bar{l},\overline{m},\overline{n},\overline{\alpha})$ coincide. Then (γ,μ) and $(\overline{\gamma},\overline{\mu})$ are congruent as framed curves.

Let $(\gamma,\mu_1,\mu_2) : I \to \mathbb{R}^3 \times \Delta_2$ be a framed curve and $\upsilon(s) = \mu_1(s) \times \mu_2(s)$. The Frenet-Serret type formula is given by

$$\begin{pmatrix} \upsilon'(s) \\ \mu_1'(s) \\ \mu_2'(s) \end{pmatrix} = \begin{pmatrix} 0 & -m(s) & -n(s) \\ m(s) & 0 & l(s) \\ n(s) & -l(s) & 0 \end{pmatrix} \begin{pmatrix} \upsilon(s) \\ \mu_1(s) \\ \mu_2(s) \end{pmatrix}.$$

Here, $l(s) = \langle \mu_1'(s), \mu_2(s)\rangle$, $m(s) = \langle \mu_1'(s), \upsilon(s)\rangle$ and $n(s) = \langle \mu_2'(s), \upsilon(s)\rangle$. Moreover, there exists a smooth mapping $\alpha : I \to \mathbb{R}$ such that:

$$\gamma'(s) = \alpha(s)\upsilon(s).$$

In addition, s_0 is a singular point of the framed curve γ if and only if $\alpha(s_0) = 0$.

In order to obtain generalized vectors for framed curves and to get a frame similar to the Frenet-Serret frame,

$$\begin{pmatrix} \overline{\mu_1}(s) \\ \overline{\mu_2}(s) \end{pmatrix} = \begin{pmatrix} \cos\theta(s) & -\sin\theta(s) \\ \sin\theta(s) & \cos\theta(s) \end{pmatrix} \begin{pmatrix} \mu_1(s) \\ \mu_2(s) \end{pmatrix}$$

is defined where $(\mu_1(s),\mu_2(s)) \in \Delta_2$ and $\theta(s)$ is a smooth function. Consequently, $(\gamma,\overline{\mu_1},\overline{\mu_2}) : I \to \mathbb{R}^3 \times \Delta_2$ also a framed curve. Let $\theta : I \to \mathbb{R}$ be a smooth function that satisfies $m(s)\sin\theta(s) = -n(s)\cos\theta(s)$. Assume that $m(s) = -p(s)\cos\theta(s), n(s) = p(s)\sin\theta(s)$, then we have:

$$\begin{aligned}
\overline{\upsilon}(s) &= \upsilon(s), \\
\upsilon'(s) &= -m(s)\mu_1(s) - n(s)\mu_2(s) = p(s)[\cos\theta(s)\mu_1(s) - \sin\theta(s)\mu_2(s)], \\
&= p(s)\overline{\mu_1}(s), \\
\overline{\mu_1}'(s) &= -(l(s) - \theta'(s))\sin\theta(s)\mu_1(s) \\
&\quad +(l(s) - \theta'(s))\cos\theta(s)\mu_2(s) + (m(s)\cos\theta(s) - n(s)\sin\theta(s))\upsilon(s), \\
&= -p(s)\upsilon(s) + (l(s) - \theta'(s))\overline{\mu_2}(s), \\
\overline{\mu_2}'(s) &= -(l(s) - \theta'(s))\cos\theta(s)\mu_1(s) \\
&\quad +(l(s) - \theta'(s))\sin\theta(s)\mu_2(s) + (m(s)\sin\theta(s) + n(s)\cos\theta(s))\upsilon(s), \\
&= -(l(s) - \theta'(s))\overline{\mu_1}(s).
\end{aligned} \tag{2.1}$$

By simple calculations, we find

$$\begin{pmatrix} \upsilon'(s) \\ \overline{\mu_1'}(s) \\ \overline{\mu_2'}(s) \end{pmatrix} = \begin{pmatrix} 0 & p(s) & 0 \\ -p(s) & 0 & q(s) \\ 0 & -q(s) & 0 \end{pmatrix} \begin{pmatrix} \upsilon(s) \\ \overline{\mu_1}(s) \\ \overline{\mu_2}(s) \end{pmatrix}.$$

We call the vectors $\upsilon(s), \overline{\mu_1}(s), \overline{\mu_2}(s)$ the generalized tangent vector, the generalized principle normal vector, and the generalized binormal vector of the framed curve, respectively, where

$$p(s) = \|\upsilon'(s)\| > 0$$

and

$$q(s) = l(s) - \theta'(s).$$

The functions $(p(s), q(s), \alpha(s))$ are referred to as the framed curvature of $\gamma(s)$ [8].

In general, the moving frame of a framed curve does not have geometric meaning. However, we can consider a moving frame with geometric meaning under a certain condition.

Definition 2.4 (Frenet-type framed curve). We say that $\gamma : I \to \mathbb{R}^3$ is a Frenet-type framed base curve if there exist a regular spherical curve $\mathcal{T} : I \to S^2$ and a function $\alpha : I \to \mathbb{R}$ such that $\gamma'(s) = \alpha(s)\mathcal{T}(s)$ for all $s \in I$. Then we call $\mathcal{T}(s)$ a unit tangent vector and $\alpha(s)$ a speed function of $\gamma(s)$ [4].

Then we have an orthonormal frame for Frenet-type framed curves such that

$$\{\mathcal{T}(s), \mathcal{N}(s), \mathcal{B}(s)\} = \{\mathcal{T}(s), \frac{\mathcal{T}'(s)}{\|\mathcal{T}'(s)\|}, \mathcal{T}(s) \times \mathcal{N}(s)\}.$$

Then we have the following Frenet-Serret type formula:

$$\begin{pmatrix} \mathcal{T}'(s) \\ \mathcal{N}'(s) \\ \mathcal{B}'(s) \end{pmatrix} = \begin{pmatrix} 0 & \kappa(s) & 0 \\ -\kappa(s) & 0 & \tau(s) \\ 0 & -\tau(s) & 0 \end{pmatrix} \begin{pmatrix} \mathcal{T}(s) \\ \mathcal{N}(s) \\ \mathcal{B}(s) \end{pmatrix},$$

where

$$\kappa(s) = \|\mathcal{T}'(s)\|, \quad \tau(s) = \frac{det(\mathcal{T}(s), \mathcal{T}'(s), \mathcal{T}''(s))}{\|\mathcal{T}'(s)\|^2}.$$

We can easily check that $(\gamma, \mathcal{N}, \mathcal{B}) : I \to \mathbb{R}^3 \times \Delta_2$ is a framed curve, so that we can apply the theory of framed curves.

3 Framed normal curves

In this section, the framed normal curves are defined, and we investigate their characterizations.

Definition 3.1. Let $(\gamma, \overline{\mu_1}, \overline{\mu_2}) : I \to \mathbb{R}^3 \times \Delta_2$ be a framed curve. We call γ a framed normal curve if its position vector γ satisfies:

$$\gamma(s) = \lambda(s)\overline{\mu_1}(s) + \varepsilon(s)\overline{\mu_2}(s)$$

for some functions $\lambda(s)$ and $\varepsilon(s)$.

Some properties of the framed normal curves are shown in the following theorems.

Theorem 3.2. Let $(\gamma, \overline{\mu_1}, \overline{\mu_2}) : I \to \mathbb{R}^3 \times \Delta_2$ be a framed curve in $\mathbb{R}^3$ with $q(s) \neq 0$. Then γ is a framed normal curve if and only if the generalized normal and binormal components of the position vector γ are respectively given by

$$\langle \gamma(s), \overline{\mu_1}(s) \rangle = -\frac{\alpha(s)}{p(s)}, \quad \langle \gamma(s), \overline{\mu_2}(s) \rangle = -\frac{1}{q(s)} \left(\frac{\alpha(s)}{p(s)} \right)'.$$

Proof. First assume that $\gamma(s)$ is a framed normal curve. Therefore, we have

$$\gamma(s) = \lambda(s)\overline{\mu_1}(s) + \varepsilon(s)\overline{\mu_2}(s). \tag{3.1}$$

By taking the derivative of (3.1) and by using the corresponding Frenet equations for framed curve, we find

$$\begin{aligned} -\lambda(s)p(s) &= \alpha(s), \\ \lambda'(s) - q(s)\varepsilon(s) &= 0, \\ \lambda(s)q(s) + \varepsilon'(s) &= 0. \end{aligned} \tag{3.2}$$

Therefore, according to first and second equations in (3.2), we have

$$\lambda(s) = -\frac{\alpha(s)}{p(s)}, \quad \varepsilon(s) = -\frac{1}{q(s)} \left(\frac{\alpha(s)}{p(s)} \right)'. \tag{3.3}$$

Consequently,

$$\gamma(s) = -\frac{\alpha(s)}{p(s)}\overline{\mu_1}(s) - \frac{1}{q(s)} \left(\frac{\alpha(s)}{p(s)} \right)' \overline{\mu_2}(s).$$

Conversely, suppose that

$$\begin{aligned} \langle \gamma(s), \overline{\mu_1}(s) \rangle &= -\frac{\alpha(s)}{p(s)}, \\ \langle \gamma(s), \overline{\mu_2}(s) \rangle &= -\frac{1}{q(s)} \left(\frac{\alpha(s)}{p(s)} \right)'. \end{aligned} \tag{3.4}$$

By taking the derivative of first equation of (3.4) with respect to s, we get

$$\langle \alpha(s)\upsilon(s), \overline{\mu_1}(s) \rangle + \langle \gamma(s), -p(s)\upsilon(s) + q(s)\overline{\mu_2}(s) \rangle = -\left(\frac{\alpha(s)}{p(s)} \right)'. \tag{3.5}$$

Therefore, by using second equation of (3.4) in (3.5), we find

$$-p(s)\langle \gamma(s), \upsilon(s) \rangle + q(s) \left(\frac{1}{q(s)} \left(\frac{\alpha(s)}{p(s)} \right)' \right) = -\left(\frac{\alpha(s)}{p(s)} \right)'.$$

It follows that $\langle \gamma(s), \upsilon(s) \rangle = 0$, which means that $\gamma(s)$ is a framed normal curve. Q.E.D.

Corollary 3.3. Let $\gamma(s)$ be a framed normal curve in $\mathbb{R}^3$. If s_0 is a singular point of the framed normal curve by according to (3.2) equations, then

$$\langle \gamma(s_0), \overline{\mu_1}(s_0)\rangle = \lambda(s_0) = 0,$$

$$\langle \gamma(s_0), \overline{\mu_2}(s_0)\rangle = \varepsilon(s_0) = \frac{\alpha' s_0}{p(s_0)q(s_0)}.$$

Theorem 3.4. Let $(\gamma, \overline{\mu_1}, \overline{\mu_2}) : I \to \mathbb{R}^3 \times \Delta_2$ be a framed curve in $\mathbb{R}^3$ with $q(s) \neq 0$. Then γ is congruent to a framed normal curve if and only if

$$-\frac{\alpha(s)q(s)}{p(s)} = \left(\frac{1}{q(s)}\left(\frac{\alpha(s)}{p(s)}\right)'\right)'. \tag{3.6}$$

Proof. Let us assume that $\gamma(s)$ is congruent to a framed normal curve. According to equation (3.4), we have relation (3.6). Conversely, assume that $\gamma(s)$ is a framed curve satisfying the relation (3.6). By applying Frenet equation for framed curve, we have

$$\frac{d}{ds}\left[\gamma(s) + \frac{\alpha(s)}{p(s)}\overline{\mu_1}(s) + \frac{1}{q(s)}\left(\frac{\alpha(s)}{p(s)}\right)'\overline{\mu_2}(s)\right] = 0.$$

Consequently, γ is a congruent to a framed normal curve. Q.E.D.

Theorem 3.5. Let $\gamma(s)$ be a framed curve with $q(s) \neq 0$. Then $\gamma(s)$ lies on sphere $S^2(r)$ for $\forall s \in I$ if and only if $\gamma(s)$ is a framed normal curve.

Proof. First assume that $\gamma(s)$ is a framed base curve lying on a sphere of radius r and center m. Therefore, we have

$$< \gamma(s) - m, \gamma(s) - m >= r^2. \tag{3.7}$$

By differentiation in s, we find that

$$2\alpha\langle \upsilon, \gamma - m\rangle = 0. \tag{3.8}$$

Let us assume that $\langle \gamma - m, \upsilon\rangle = h$ for every $s \in I$. Therefore, if $\alpha h = 0$ for each $s \in I$, it is either $\alpha = 0$ or $h = 0$. If $\alpha(s) = 0$ for all $s \in I$, then γ is a point [3]. Consequently, since γ is a framed spherical curve, $h = 0$ for every $s \in I$. Even if $\alpha(s_0) = 0$ for $s = s_0$ singular point, $h(s_0) = 0$ will be provided. Therefore, we have that

$$\langle \upsilon, \gamma - m\rangle = 0, \tag{3.9}$$

for each $s \in I$. By differentiation in s of equations (3.9) two times and applying Frenet formulae, we find

$$\gamma(s) - m = -\frac{\alpha(s)}{p(s)}\overline{\mu_1}(s) - \frac{1}{q(s)}\left(\frac{\alpha(s)}{p(s)}\right)'\overline{\mu_2}(s).$$

Consequently, γ is a framed normal curve. Conversely, suppose that $\gamma(s)$ is a framed normal curve. According to Theorem 3.4, we have

$$-\frac{\alpha(s)q(s)}{p(s)} = \left(\frac{1}{q(s)}\left(\frac{\alpha(s)}{p(s)}\right)'\right)'. \tag{3.10}$$

Let us consider the vector

$$m = \gamma(s) + \frac{\alpha(s)}{p(s)}\overline{\mu_1}(s) + \frac{1}{q(s)}\left(\frac{\alpha(s)}{p(s)}\right)^{'}\overline{\mu_2}(s).$$

Differentiating this equation with respect to s, we get $m^{'} = 0$. Consequently, $m = constant$. Since (3.10) is the differential of the equation

$$\left(\frac{\alpha(s)}{p(s)}\right)^2 + \left(\frac{1}{q(s)}\left(\frac{\alpha(s)}{p(s)}\right)^{'}\right)^2 = c = constant.$$

Therefore, we have

$$\langle \gamma - m, \gamma - m\rangle = \left(\frac{\alpha(s)}{p(s)}\right)^2 + \left(\frac{1}{q(s)}\left(\frac{\alpha(s)}{p(s)}\right)^{'}\right)^2 = c = constant.$$

We may take $c = r^2$. Finally, it follows that image of the framed base curve $\gamma(s)$ lies on a sphere of radius r.

Q.E.D.

Theorem 3.6. Let $(\gamma, \overline{\mu_1}, \overline{\mu_2}) : I \to \mathbb{R}^3 \times \Delta_2$ be a framed curve in $\mathbb{R}^3$ with $q(s) \neq 0$. $\gamma(s)$ is a framed normal curve if and only if the following statements are satisfied:
i.) The framed curvatures $p(s), q(s)$ and $\alpha(s)$ satisfy the following equality

$$\frac{\alpha(s)}{p(s)} = c_1 \cos\left(\int q(s)ds\right) + c_2 \sin\left(\int q(s)ds\right)$$

for some constant $c_1, c_2 \in \mathbb{R}$
ii.) The generalized principal normal and generalized binormal component of the position vector of the framed curve are given respectively by

$$\langle \gamma(s), \overline{\mu_1}(s)\rangle = -c_1 \cos\left(\int q(s)ds\right) - c_2 \sin\left(\int q(s)ds\right),$$

$$\langle \gamma(s), \overline{\mu_2}(s)\rangle = -c_1 \sin\left(\int q(s)ds\right) + c_2 \cos\left(\int q(s)ds\right).$$

Conversely if $\gamma(s)$ is framed curve in $\mathbb{R}^3$, the framed curvature $q(s) \neq 0$ for each $s \in I$ and one of the statements (i) and (ii) hold, then γ is a framed normal curve or congruent to a framed normal curve.

Proof. First assume that $\gamma(s)$ is a framed normal curve. According to equation (3.6), we have

$$-\frac{\alpha(s)q(s)}{p(s)} = \left(\frac{1}{q(s)}\left(\frac{\alpha(s)}{p(s)}\right)^{'}\right)^{'}.$$

Putting $t(s) = \dfrac{1}{p(s)}$ and $z(s) = \dfrac{1}{q(s)}$, equation (3.6) can be written as

$$-\frac{\alpha(s)t(s)}{z(s)} = \left(z(s)[\alpha(s)t(s)]'\right)'$$

If we change variables in the last equation as $x = \int \frac{1}{z(s)}ds$, then we have

$$\frac{d^2(\alpha t)}{dx^2} + \alpha t = 0 \tag{3.11}$$

The solution of (3.11) differential equation is

$$\alpha(s)t(s) = c_1 \cos x + c_2 \sin x$$

where $c_1, c_2 \in \mathbb{R}$. Consequently,

$$\frac{\alpha(s)}{p(s)} = c_1 \cos\left(\int q(s)ds\right) + c_2 \sin\left(\int q(s)ds\right) \tag{3.12}$$

Therefore, we have proved statement (i). According to equation (3.3) and (3.12), we get

$$\lambda(s) = -c_1 \cos\left(\int q(s)ds\right) - c_2 \sin\left(\int q(s)ds\right)$$

$$\varepsilon(s) = -c_1 \sin\left(\int q(s)ds\right) + c_2 \cos\left(\int q(s)ds\right).$$

Consequently, the framed normal curve is denote by

$$\begin{aligned}\gamma(s) = &\left[-c_1 \cos\left(\textstyle\int q(s)ds\right) - c_2 \sin\left(\textstyle\int q(s)ds\right)\right] \overline{\mu_1}(s) + \\ &\left[-c_1 \sin\left(\textstyle\int q(s)ds\right) + c_2 \cos\left(\textstyle\int q(s)ds\right)\right] \overline{\mu_2}(s).\end{aligned} \tag{3.13}$$

Therefore, from (3.13), we have

$$\langle \gamma(s), \gamma(s) \rangle = c_1^2 + c_2^2, \tag{3.14}$$

$$\langle \gamma(s), \overline{\mu_1}(s) \rangle = -c_1 \cos\left(\int q(s)ds\right) - c_2 \sin\left(\int q(s)ds\right), \tag{3.15}$$

$$\langle \gamma(s), \overline{\mu_1}(s) \rangle = -c_1 \sin\left(\int q(s)ds\right) + c_2 \cos\left(\int q(s)ds\right). \tag{3.16}$$

Therefore, we have proved statement (ii). Conversely, let us suppose that statement (i) holds. Therefore, we have

$$\frac{\alpha(s)}{p(s)} = c_1 \cos\left(\int q(s)ds\right) + c_2 \sin\left(\int q(s)ds\right).$$

Differentiating last equation with respect to s, we get

$$-\frac{\alpha(s)q(s)}{p(s)} = \left(\frac{1}{q(s)}\left(\frac{\alpha(s)}{p(s)}\right)'\right)'.$$

By applying Frenet equations (1) for framed curves, we find

$$\frac{d}{ds}\left[\gamma(s) + \frac{\alpha(s)}{p(s)}\overline{\mu_1}(s) + \frac{1}{q(s)}\left(\frac{\alpha(s)}{p(s)}\right)'\overline{\mu_2}(s)\right] = 0.$$

Consequently, γ is a framed normal curve. Next, let us suppose that statement (ii) holds. Therefore, we have

$$\langle\gamma(s), \overline{\mu_1}(s)\rangle = -c_1\cos\left(\int q(s)ds\right) - c_2\sin\left(\int q(s)ds\right), \tag{3.17}$$

$$\langle\gamma(s), \overline{\mu_2}(s)\rangle = -c_1\sin\left(\int q(s)ds\right) + c_2\cos\left(\int q(s)ds\right). \tag{3.18}$$

By differentiation in s of (3.17) and using (3.18), we get

$$\langle\gamma(s), \upsilon(s)\rangle = 0. \tag{3.19}$$

Consequently, by according to (3.19), γ is a framed normal curve

Q.E.D.

Corollary 3.7. Let $\gamma(s)$ be a framed normal curve. If s_0 is a singular point of the framed normal curve, then

$$c_1\cos\left(\int q(s)ds\right) = -c_2\sin\left(\int q(s)ds\right)$$

for $s_0 \in I$.

4 Frenet-type framed normal curves

Definition 4.1. Let $(\gamma, \mathcal{N}, \mathcal{B}) : I \to \mathbb{R}^3 \times \Delta_2$ be a framed curve. We call γ a framed normal curve if its position vector γ satisfies:

$$\gamma(s) = \lambda(s)\mathcal{N}(s) + \varepsilon(s)\mathcal{B}(s)$$

for some functions $\lambda(s)$ and $\varepsilon(s)$.

Some properties of the Frenet-type framed normal curves are shown in the following theorems. The proofs of the Theorems are similar to the Theorem (3.3) and Theorem (3.5).

Theorem 4.2. Let $(\gamma, \mathcal{N}, \mathcal{B}) : I \to \mathbb{R}^3 \times \Delta_2$ be a framed curve in $\mathbb{R}^3$ with framed curvatures $\kappa(s) > 0$, $\tau(s) \neq 0$. Then γ is a framed normal curve if and only if the normal and binormal components of the position vector γ are respectively given by

$$\langle\gamma(s), \mathcal{N}(s)\rangle = -\frac{\alpha(s)}{\kappa(s)}, \quad \langle\gamma(s), \mathcal{B}(s)\rangle = -\frac{1}{\tau(s)}\left(\frac{\alpha(s)}{\kappa(s)}\right)'.$$

Theorem 4.3. Let $(\gamma, \mathcal{N}, \mathcal{B}) : I \to \mathbb{R}^3 \times \Delta_2$ be a framed curve in $\mathbb{R}^3$ with framed curvatures $\kappa(s) > 0$, $\tau(s) \neq 0$. Then γ is congruent to a framed normal curve if and only if

$$-\frac{\alpha(s)\tau(s)}{\kappa(s)} = \left(\frac{1}{\tau(s)}\left(\frac{\alpha(s)}{\kappa(s)}\right)'\right)'. \tag{4.1}$$

Example 4.4. (The spherical nephroid ([4]). The spherical nephroid

$$\gamma : [0, 2\pi) \to S^2 \subset \mathbb{R}^3$$

is defined by

$$\gamma(s) = \left(\frac{3}{4}\cos s - \frac{1}{4}\cos 3s, \frac{3}{4}\sin s - \frac{1}{4}\sin 3s, \frac{\sqrt{3}}{2}\cos s\right). \tag{4.2}$$

See Fig.1. Then

$$\mathcal{T}(s) = \frac{1}{2}\left(\sqrt{3}\cos 2s, \sqrt{3}\sin 2s, -1\right) \tag{4.3}$$

gives the unit tangent vector and $\alpha(s) = \sqrt{3}\sin s$. By a calculation, we get

$$\mathcal{N}(s) = (-\sin 2s, \cos 2s, 0)\,, \tag{4.4}$$

$$\mathcal{B}(s) = \frac{1}{2}\left(\cos 2s, \sin 2s, \sqrt{3}\right). \tag{4.5}$$

Moreover, we find,

$$\kappa(s) = \|\mathcal{T}'(s)\| = \sqrt{3}$$

$$\tau(s) = \frac{det(\mathcal{T}(s), \mathcal{T}'(s), \mathcal{T}''(s))}{\|\mathcal{T}'(s)\|^2} = -1.$$

According to equations (3.1), (3.3) and (3.4), we get

$$\left(\frac{3}{4}\cos s - \frac{1}{4}\cos 3s, \frac{3}{4}\sin s - \frac{1}{4}\sin 3s, \frac{\sqrt{3}}{2}\cos s\right) = \lambda(s)\left(-\sin 2s, \cos 2s, 0\right) + \varepsilon(s)\left(\frac{\cos 2s}{2}, \frac{\sin 2s}{2}, \frac{\sqrt{3}}{2}\right)$$

Therefore, we find $\lambda(s) = -\sin s$ and $\varepsilon(s) = \cos s$. Moreover, $\lambda(s_0) = 0$ in singular point of γ. Moreover, according to theorem 4.2, we have

$$\langle \gamma(s), \mathcal{N}(s) \rangle = -\sin s \tag{4.6}$$

and

$$-\frac{\alpha(s)}{\kappa(s)} = -\sin s, \tag{4.7}$$

$$\langle \gamma(s), \mathcal{B}(s) \rangle = \cos s \tag{4.8}$$

and

$$-\frac{1}{\tau(s)}\left(\frac{\alpha(s)}{\kappa(s)}\right)' = \cos s. \tag{4.9}$$

Consequently, γ is a framed normal curve.

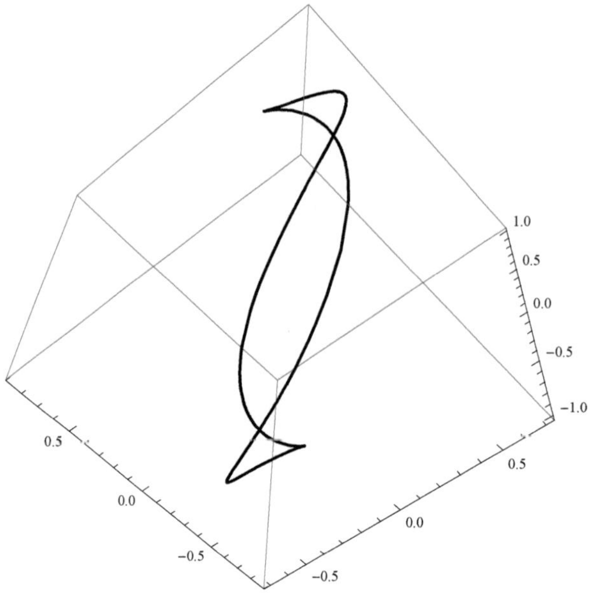

FIGURE 1. Spherical Nephroid is a framed normal curve.

References

[1] B. Y. Chen, *When does the position of a space curve always lie in its rectifying plane?*. Amer. Math. Mounthly. (2003); 110: 147-152.

[2] T. Fukunaga, M. Takahashi, *Existence conditions of framed curves for smooth curves.* Journal of Geometry. (2017); 108: 763-774.

[3] S. Honda, M. Takahashi, *Framed curves in the Euclidean space.* Advances in Geometry. (2016); 16(3): 265-276.

[4] S. Honda, *Rectifying developable surfaces of framed base curves and framed helices.* Advanced Studies in Pure Mathematics. (2018); 78: 273-292.

[5] K. İlarslan, *Spacelike normal curves in Minkowski space E_1^3.* Turkish Journal of Mathematics. (2005); 29: 53-63.

[6] K. İlarslan, E. Nesovic, *Spacelike and timelike normal curves in Minkowski space-time.* Publications de l'Institut Mathmatique. (2009); 85(99): 111-118.

[7] M. Önder, H. H. Uğurlu, *Normal and spherical curves in Dual space D^3.* Mediterranean Journal of Mathematics. (2013); 10: 1527-1537.

[8] Y. Wang, D. Pei, R. Gao, *Generic properties of framed rectifying curves.* Mathematics.(2019); 7,37.

[9] Y. C. Wong, *A global formulation of the condition for a curve to lie in a sphere*. Monatsh.Math. (1963),67: 363-365.

[10] Y. C. Wong, *On an explicit characterization of spherical curves*. Proc. Amer. Math. Soc. (1972); 34: 239-242.

DOI 10.2478/9788395793882-004

Geometry of coupled dispersionless equations by using bishop frames

Kemal Eren

Department of Mathematics, Faculty of Arts and Sciences,Sakarya University, Sakarya, Turkey
E-mail: kemaleren52@gmail.com

Abstract

In this study, we investigate the equations of the coupled dispersionless and complex coupled dispersionless by an alternative point of view to avoid insufficiency occurs for sudden orientation changes when the curve straightens out momentarily by considering a moving coordinate frame on a space curve based on the concept of parallel transport. Firstly, we study the correlation between the type-1 Bishop frame of a space curve and the coupled dispersionless equation: then we discover the correlation between the type-2 Bishop frame and the complex coupled dispersionless equations. Finally, we give the integrability conditions of the obtained differential equations by finding the Lax pair.

2010 Mathematics Subject Classification. **53A04** 14H70
Keywords. Bishop frames, coupled dispersionless equations, Lax pairs.

1 Introduction

Most of the integrable systems arise as a quasi-classical limit of ordinary integrable systems with a dispersion term. There are however important examples of integrable systems which are referred to as dispersionless, not in this sense, but because the dispersion term is absent. The class of integrable partial differential equations called dispersionless integrable systems have attracted much interest in view of their wide range of applications in various fields of mathematics and physics.

The development process of the dispersionless systems can be seen in the table below;

TABLE 1. Dispersionless Equations

Type of Dispersionless Equations	Abbreviations	Equation systems	References
Coupled Dispersionless	CD	$\rho_s + uu_y = 0,$ $u_{ys} = \rho u,$	[1, 2]
Generalized Coupled Dispersionless	GCD	$\rho_s + \frac{(uv)_y}{2} = 0,$ $u_{ys} = \rho u,$ $v_{ys} = \rho v,$	[3]
Complex Couple Dispersionless	CCD	$\rho_s + \frac{1}{2}\left(\lvert u\rvert^2\right)_y = 0,$ $u_{ys} = \rho u.$	[4]

These equation systems have been generalized by many researchers in many different aspects and proposed as models for describing physical phenomena [5]-[16]. For instance, Shen, et al. have studied the CCD equation and the complex short pulse equation from the algebraic and geometric point of view and they have exposed the link of the motions of space curves to the real and complex

Tbilisi Mathematical Journal Special Issue (IECMSA-2019), pp. 39–48.
Tbilisi Centre for Mathematical Sciences.

Received by the editors: 04 October 2019.
Accepted for publication: 06 January 2020.

CD equations and short pulse equations via hodograph transformations by using the Serret-Frenet frame of the curves in [17].

On the other hand, the theory of curves is primarily formed by the Serret-Frenet formulas which describe the kinematic properties of a particle moving along a differentiable curve or the geometric properties of the curve itself irrespective of any motion. A great deal of research has been conducted on the Frenet frame formalism of space curves. However, it is known that it is not possible to calculate the Frenet frame for straight-line segments and at points where the second derivative vanishes. So, there is an alternative way to define the moving reference frame that was introduced by L.R. Bishop in 1975 by means of parallel vector fields and so-called as alternative or parallel frame of the curves [18]. Recently, some researchers called this frame as Bishop (parallel transport) frame and this frame based studies have been intensively treated, see [19, 20, 21, 22]. In these regards, we study the link of the coupled dispersionless equations and short pulse equations with space curves according to Bishop frames.

2 Preliminaries

Let $\gamma = \gamma(s)$ be a regular unit speed curve in Euclidean 3-space. If T, N, and B denote the tangent, principal normal and binormal unit vectors at $\gamma(s)$ point of the curve γ, respectively. Then the Frenet formulas are given

$$\begin{bmatrix} T' \\ N' \\ B' \end{bmatrix} = \begin{bmatrix} 0 & \kappa & 0 \\ -\kappa & 0 & \tau \\ 0 & -\tau & 0 \end{bmatrix} \begin{bmatrix} T \\ N \\ B \end{bmatrix}$$

where $\kappa = \|T'\|$ is curvature and $\tau = -\langle N, B' \rangle$ is torsion of the curve γ. This classical frame $\{T, N, B\}$ provides to illustrate a lot of properties of a curve; however, it is not defined for all points along every curve. The alternative way of defining a moving reference frame is Bishop frame. The Bishop frame provides significant advantages for curves whose second derivative is zero at some points according to the Frenet frame. This frame is constructed in two different ways as the type-1 Bishop frame $\{T, N_1, N_2\}$ and the type-2 Bishop frame $\{M_1, M_2, B\}$. Let's give the relations between the Frenet frame and the Bishop frames.

The type-1 Bishop frame $\{T, N_1, N_2\}$ is based on the model where T is the tangent vector and any convenient arbitrary basis is $\{N_1, N_2\}$ for the normal plane perpendicular to the tangent vector. Then the derivative equations of type-1 Bishop frame are given by

$$\begin{bmatrix} T' \\ {N_1}' \\ {N_2}' \end{bmatrix} = \begin{bmatrix} 0 & k_1 & k_2 \\ -k_1 & 0 & 0 \\ -k_2 & 0 & 0 \end{bmatrix} \begin{bmatrix} T \\ N_1 \\ N_2 \end{bmatrix} \tag{2.1}$$

where $\langle T, T \rangle = \langle N_1, N_1 \rangle = \langle N_2, N_2 \rangle = 1$, $\langle T, N_1 \rangle = \langle T, N_2 \rangle = \langle N_1, N_2 \rangle = 0$. Also k_1 and k_2 are called the curvatures of the curve according to the type-1 Bishop frame. Let θ be the angle between the vectors N_1 and N of γ, then the relation between the Frenet frame and the type-1 Bishop frame is

$$\begin{bmatrix} T \\ N_1 \\ N_2 \end{bmatrix} = \begin{bmatrix} 1 & 0 & 0 \\ 0 & \cos\theta & -\sin\theta \\ 0 & \sin\theta & \cos\theta \end{bmatrix} \begin{bmatrix} T \\ N \\ B \end{bmatrix}.$$

Moreover, the relations between the curvatures of curve according to Frenet and type-1 Bishop frames are given

$$\kappa = \sqrt{k_1^2 + k_2^2},\ \tau = \frac{d\theta}{ds}$$

or

$$k_1 = \kappa\cos\theta, k_2 = \kappa\sin\theta, \theta = \arctan\left(\frac{k_2}{k_1}\right) = \int \tau ds.$$

On the other hand, let us denote type-2 Bishop frame of a curve γ by $\{M_1, M_2, B\}$ then the derivative equations of type-2 Bishop frame are given as

$$\begin{bmatrix} M_1{}' \\ M_2{}' \\ B' \end{bmatrix} = \begin{bmatrix} 0 & 0 & -\varepsilon_1 \\ 0 & 0 & -\varepsilon_2 \\ \varepsilon_1 & \varepsilon_2 & 0 \end{bmatrix} \begin{bmatrix} M_1 \\ M_2 \\ B \end{bmatrix} \tag{2.2}$$

such that $\langle M_1, M_1\rangle = \langle M_2, M_2\rangle = \langle B, B\rangle = 1$, $\langle M_1, M_2\rangle = \langle M_1, B\rangle = \langle M_2, B\rangle = 0$. Here ε_1 and ε_2 denote the curvatures of the curve according to the type-2 Bishop frame. Let φ be the angle between the vectors N_2 and N, then there is the relation between the Frenet frame and the type-2 Bishop frame as

$$\begin{bmatrix} M_1 \\ M_2 \\ B \end{bmatrix} = \begin{bmatrix} \sin\varphi & \cos\varphi & 0 \\ -\cos\varphi & \sin\varphi & 0 \\ 0 & 0 & 1 \end{bmatrix} \begin{bmatrix} T \\ N \\ B \end{bmatrix}.$$

The relation between the curvatures of curve according to the Frenet frame and Bishop-2 frame is

$$\tau = \sqrt{\varepsilon_1^2 + \varepsilon_2^2},\quad \kappa = \frac{d\varphi}{ds}$$

or

$$\varepsilon_1 = -\tau\cos\varphi, \varepsilon_2 = -\tau\sin\varphi, \varphi = \arctan\left(\frac{\varepsilon_2}{\varepsilon_1}\right) = \int \kappa ds.$$

3 The link of the coupled dispersionless equation with space curves according to Bishop frames

In this part of the study, we investigate the correlation between the Bishop frames of space curves and the equations of CD. Let's assume that

$$\gamma(y, s) : [0, l]\, x\, [0, S] \to E^3$$

is a family of space curves, where $y \in [0, l]$ is the arc-length parameter and s represents the time.

Theorem 3.1. Let $\gamma(y, s)$ be a family of space curves, then the following statement provides the CD equation

$$\begin{bmatrix} T \\ N_1 \\ N_2 \end{bmatrix}_s = \begin{bmatrix} 0 & -c^{-1} & 0 \\ c^{-1} & 0 & -u \\ 0 & u & 0 \end{bmatrix} \begin{bmatrix} T \\ N_1 \\ N_2 \end{bmatrix}, \tag{3.1}$$

where $\{T, N_1, N_2\}$ is the type-1 Bishop frame, c is a non-zero constant, u is a real function.

Proof. We obtain time evolution for the orthogonal frame $\{T, N_1, N_2\}$ in matrix form as

$$\begin{bmatrix} T \\ N_1 \\ N_2 \end{bmatrix}_s = \begin{bmatrix} 0 & \alpha & \beta \\ -\alpha & 0 & \delta \\ -\beta & -\delta & 0 \end{bmatrix} \begin{bmatrix} T \\ N_1 \\ N_2 \end{bmatrix} \tag{3.2}$$

where α, β and δ are functions of y and s. Also, from the type-1 Bishop formulae given by (2.1), we write

$$\begin{bmatrix} T \\ N_1 \\ N_2 \end{bmatrix}_y = \begin{bmatrix} 0 & k_1 & k_2 \\ -k_1 & 0 & 0 \\ -k_2 & 0 & 0 \end{bmatrix} \begin{bmatrix} T \\ N_1 \\ N_2 \end{bmatrix}.$$

Thus, under favorable conditions $T_{sy} = T_{ys}, N_{1,sy} = N_{1,ys}, N_{2,sy} = N_{2,ys}$ we have

$$\alpha_y = k_{1,s} - k_2\delta, \tag{3.3}$$

$$\beta_y = k_1\delta + k_{2,s}, \tag{3.4}$$

$$\delta_y = -k_1\beta + k_2\alpha. \tag{3.5}$$

By the hypothesis $\alpha = -c^{-1}$, $\beta = 0$ and $\delta = -u$ are satisfied and then the equations (3.3)-(3.5) become

$$0 = k_{1,s} + k_2 u, \tag{3.6}$$

$$0 = -k_1 u + k_{2,s}, \tag{3.7}$$

$$u_y = k_2 c^{-1}, \tag{3.8}$$

respectively. So, from the equation (3.8) we easily see that

$$k_2 = c u_y. \tag{3.9}$$

By substituting the equation (3.9) into the equation (3.6), we find

$$k_1 = c\rho \tag{3.10}$$

under consideration $\rho_s = -u u_y$. By substituting the equations (3.9) and (3.10) into the equations (3.6) and (3.7), respectively, we have

$$\rho_s + u u_y = 0 \tag{3.11}$$

and

$$u_{ys} = \rho u. \tag{3.12}$$

As a result, the equations (3.11) and (3.12) express the CD equation and this completes the proof.

Q.E.D.

Corollary 3.2. The CD equation corresponds to the set

$$\{k_1, k_2, \alpha, \beta, \delta\} = \{c\rho, cu_y, -c^{-1}, 0, -u\}.$$

In 1968 P.D. Lax introduced the Lax equations providing the integrability of nonlinear differential equations [2, 23]. Now, let's give the Lax pair which provides integrability of CD equation by the following theorem.

Theorem 3.3. Let's take a function $\psi = \psi(y, s)$ with so(3) value, then Lax pair of the CD equation is

$$\psi_y = U\psi, \psi_s = V\psi \tag{3.13}$$

such that $U = -k_1 e_3 + k_2 e_2$ and $V = -\alpha e_3 - \delta e_1$, where $k_1 = \langle T_y, N_1\rangle$, $k_2 = \langle T_y, N_2\rangle$, $\alpha = \langle T_s, N_1\rangle$, $\delta = \langle N_{1s}, N_2\rangle$ and $\{T, N_1, N_2\}$ is the type-1 Bishop frame of the space curve $\gamma(y, s)$.

Proof. The Lax pair of the CD equation is

$$U = -i\lambda \begin{pmatrix} \rho & u_y \\ u_y & -\rho \end{pmatrix}, V = \begin{pmatrix} \frac{i}{4\lambda} & \frac{-u}{2} \\ \frac{u}{2} & -\frac{i}{4\lambda} \end{pmatrix}. \tag{3.14}$$

Also, the compatibility condition $U_y - V_s + UV - VU = 0$ satisfies the CD equation [3]. The basis of su(2) and so(3) are

$$e_1 = \frac{1}{2i}\begin{pmatrix} 0 & 1 \\ 1 & 0 \end{pmatrix}, e_2 = \frac{1}{2i}\begin{pmatrix} 0 & -i \\ i & 0 \end{pmatrix}, e_3 = \frac{1}{2i}\begin{pmatrix} 1 & 0 \\ 0 & -1 \end{pmatrix} \tag{3.15}$$

and

$$L_1 = \begin{pmatrix} 0 & 0 & 0 \\ 0 & 0 & -1 \\ 0 & 1 & 0 \end{pmatrix}, L_2 = \begin{pmatrix} 0 & 0 & 1 \\ 0 & 0 & 0 \\ -1 & 0 & 0 \end{pmatrix}, L_3 = \begin{pmatrix} 0 & -1 & 0 \\ 1 & 0 & 0 \\ 0 & 0 & 0 \end{pmatrix}, \tag{3.16}$$

respectively [24] and there is an isomorphism $L_j \to e_j$, $(j = 1, 2, 3)$ between the Lie algebras su(2) and so(3). Under this isomorphism, the CD equation provides the equation (3.13). From the corollary 3.2 and hypothesis, the functions U and V are found as

$$U = -k_1 e_3 - k_2 e_1 = -c\rho e_3 - cu_y e_1 = -i\lambda \begin{pmatrix} \rho & u_y \\ u_y & -\rho \end{pmatrix},$$
$$V = -\alpha e_3 - \delta e_2 = c^{-1} e_3 + u e_2 = \begin{pmatrix} \frac{i}{2\lambda} & \frac{-u}{2} \\ \frac{u}{2} & \frac{-i}{2\lambda} \end{pmatrix},$$

where $c = -2\lambda$. They provide a Lax pair of the CD equation as it is desired. Q.E.D.

Let us give the geometric interpretation of the conserved quantity of the CD equation by the following theorem.

Theorem 3.4. The conserved quantity of the CD equation is constant,

$$I = \rho^2 + u_y^2$$

where $\rho = \frac{k_1}{c}$, $u_y = \frac{k_2}{c}$, k_1 and k_2 are the type-1 Bishop curvatures of the space curve $\gamma(y, s)$.

Proof. Considering the equations (3.6) and (3.7), we find

$$\frac{d}{ds}\left({k_1}^2 + {k_2}^2\right) = 2k_1 k_{1,s} + 2k_2 k_{2,s} = 2k_1\left(-uk_2\right) + 2k_2\left(uk_1\right) = 0.$$

On the other hand, from the equation (3.9) and (3.10), we get

$$\frac{d}{ds}\left({k_1}^2 + {k_2}^2\right) = \frac{d}{ds}\left(c^2\rho^2 + (cu_y)^2\right) = c^2\frac{d}{ds}\left(\rho^2 + u_y^2\right).$$

As a result, for $c \neq 0$, we find out

$$\frac{d}{ds}\left(\rho^2 + u_y^2\right) = 0.$$

Hence, we can easily see that the conserved quantity of the CD equation is constant. Q.E.D.

Corollary 3.5. If u is a real function of y and s then the CD equation is obtained by condition (3.1).

Theorem 3.6. Let $\gamma(y, s)$ be a space curve family, then the following statements provide the CCD equation.

$$\begin{bmatrix} M_1 \\ M_2 \\ B \end{bmatrix}_s = \begin{bmatrix} 0 & -c^{-1} & u_1 \\ c^{-1} & 0 & -u_2 \\ -u_1 & u_2 & 0 \end{bmatrix}\begin{bmatrix} M_1 \\ M_2 \\ B \end{bmatrix} \tag{3.17}$$

where $\{M_1, M_2, B\}$ is the type-2 Bishop frame, $c \neq 0$ is constant, $u = u_1 + iu_2$ is a complex function.

Proof. We obtained time evolution for the orthogonal frame $\{M_1, M_2, B\}$ in matrix form as

$$\begin{bmatrix} M_1 \\ M_2 \\ B \end{bmatrix}_s = \begin{bmatrix} 0 & \eta & \mu \\ -\eta & 0 & \omega \\ -\eta & -\omega & 0 \end{bmatrix}\begin{bmatrix} M_1 \\ M_2 \\ B \end{bmatrix} \tag{3.18}$$

where η, μ and ω are functions of y and s. Also from the type-2 Bishop formulae given by (2.2), we write

$$\begin{bmatrix} M_1 \\ M_2 \\ B \end{bmatrix}_y = \begin{bmatrix} 0 & 0 & -\varepsilon_1 \\ 0 & 0 & -\varepsilon_2 \\ \varepsilon_1 & \varepsilon_2 & 0 \end{bmatrix}\begin{bmatrix} M_1 \\ M_2 \\ B \end{bmatrix}.$$

Thus, under favorable conditions $T_{sy} = T_{ys}$, $M_{1,sy} = M_{1,ys}$, $M_{2,sy} = M_{2,ys}$ we have

$$\eta_y = \varepsilon_1\omega - \varepsilon_2\mu, \tag{3.19}$$

$$\mu_y = \varepsilon_2\eta - \varepsilon_{1,s}, \tag{3.20}$$

$$\omega_y = -\varepsilon_{2,s} - \varepsilon_1 \eta. \tag{3.21}$$

From the equations (3.20) and (3.21), we obtain

$$(\varepsilon_1 - i\varepsilon_2)_s = i\eta(\varepsilon_1 - i\varepsilon_2) - (\mu - i\omega)_y. \tag{3.22}$$

Let us assume that $\varepsilon_1 - i\varepsilon_2 = icu_y, \eta = -c^{-1}, \mu - i\omega = u$ in this equation, then we have

$$\varepsilon_1 = -cu_{2y}, \varepsilon_2 = -cu_{1y} \tag{3.23}$$

and

$$\eta = -c^{-1}, \mu = u_1, \omega = -u_2. \tag{3.24}$$

By substitution the equations (3.23) and (3.24) into the equations (3.22), we have

$$u_{ys} = 0 \tag{3.25}$$

under consideration $\rho = 0$. By substituting the equations (3.23) and (3.24) into the equations (3.19), we have

$$u_1 u_{1y} + u_2 u_{2y} = 0. \tag{3.26}$$

On the other hand, we find

$$\rho_s + \frac{1}{2}\left(|u|^2\right)_y = \frac{1}{2}\left(uu_y^* + u_y u^*\right) = u_1 u_{1y} + u_2 u_{2y}. \tag{3.27}$$

From the equations (3.26) and (3.27), we obtain

$$\rho_s + \frac{1}{2}\left(|u|^2\right)_y = 0. \tag{3.28}$$

As a result, from (3.25) and (3.28) we obtained the CCD equation Thus, the theorem is proved.

Q.E.D.

Corollary 3.7. The CCD equation corresponds to the set

$$\{\varepsilon_1, \varepsilon_2, \eta, \mu, \omega\} = \left\{-cu_{2y}, -cu_{1y}, -c^{-1}, u_1, u_2\right\}.$$

Let's give the Lax pair which provides integrability of CCD equation by the theorem.

Theorem 3.8. Let's take the function $\psi(y, s)$ with $so(3)$ value, then Lax pair of the CCD equation is

$$\psi_y = P\psi, \psi_s = Q\psi \tag{3.29}$$

such that $P = -\varepsilon_1 e_1 + \varepsilon_2 e_1$ and $Q = -\eta e_3 + \mu e_2 - \omega e_1$, where $\varepsilon_1 = -\langle M_1, B\rangle$, $\varepsilon_2 = -\langle M_2, B\rangle$, $\eta = \langle M_1, M_2\rangle$, $\mu = \langle M_1, B\rangle$, $\omega = \langle M_2, B\rangle$ and $\{M_1, M_2, B\}$ is the type-2 Bishop frame of the space curve $\gamma(y, s)$.

Proof. The Lax pair of the CCD equation is given by [4] as follows

$$P=-i\lambda\begin{pmatrix}0 & u_y\\ u_y^* & 0\end{pmatrix}, Q=\begin{pmatrix}\frac{i}{4\lambda} & \frac{-u}{2}\\ \frac{u^*}{2} & -\frac{i}{4\lambda}\end{pmatrix}. \tag{3.30}$$

Also, the compatibility condition $P_y - Q_s + PQ - QP = 0$ satisfies the CCD equation and u^* is the complex conjugate of u. From the equations (3.15) and (3.16), the basis of su (2) and so (3) are $\{e_1, e_2, e_3\}$ and $\{L_1, L_2, L_3\}$, respectively. Let $L_j \to e_j$, $(j = 1, 2, 3)$ be an isomorphism between the Lie algebras su (2) and so (3), Under this isomorphism, the CCD equation provides the equation (3.29). According to type-2 Bishop frame, P and Q functions are found as

$$\begin{aligned}P&=-\varepsilon_1 e_1+\varepsilon_2 e_1=\tfrac{1}{2}\begin{pmatrix}0 & \varepsilon_1-i\varepsilon_2\\ -\varepsilon_1-i\varepsilon_2 & 0\end{pmatrix}=-i\lambda\begin{pmatrix}0 & u_y\\ u_y^* & 0\end{pmatrix}\\ Q&=-\eta e_3+\mu e_2-\omega e_1=\tfrac{1}{2}\begin{pmatrix}i\eta & -\mu+i\omega\\ \mu+i\omega & -i\eta\end{pmatrix}=\begin{pmatrix}\frac{i}{4\lambda} & \frac{-u}{2}\\ \frac{u^*}{2} & \frac{-i}{4\lambda}\end{pmatrix}\end{aligned}, \tag{3.31}$$

where $c = -2\lambda$, P and Q provide $P_y - Q_s + PQ - QP = 0$. Q.E.D.

Let us give the geometric interpretation of the conserved quantity of the CCD equation with the following theorem:

Theorem 3.9. The conserved quantity of the CCD equation is constant,

$$I=\rho^2+|u_y|^2$$

where $u_{2y} = -\frac{\varepsilon_1}{c}$, $u_{1y} = -\frac{\varepsilon_2}{c}$, ε_1 and ε_2 are the type-2 Bishop curvature of the space curve family $\gamma(y, s)$.

Proof. Considering the equations (3.20), (3.21), (3.23) and (3.24), we find

$$\frac{d}{ds}\left({k_1}^2+{k_2}^2\right)=2k_1k_{1,s}+2k_2k_{2,s}=2cu_{1y}u_{2y}-2cu_{1y}u_{2y}=0.$$

On the other hand, from the equation (3.23), we get

$$\frac{d}{ds}\left({k_1}^2+{k_2}^2\right)=\frac{d}{ds}\left(\left(-cu_{2y}\right)^2+\left(-cu_{1y}\right)^2\right)=c^2\frac{d}{ds}\left(\left(u_{2y}\right)^2+\left(u_{1y}\right)^2\right)=c^2\frac{d}{ds}\left(|u|^2\right)_y.$$

As a result, when the last two equations are equalized for $c \neq 0$, we get

$$\frac{d}{ds}\left(|u|^2\right)_y=\frac{d}{ds}\left(\rho^2+|u|^2\right)_y=0.$$

Hence, we can easily see that the conserved quantity of the CCD equation is constant. Q.E.D.

Corollary 3.10. If $u = u_1 + iu_2$ is a complex function of y and s then the CCD equation obtained by condition (3.17).

Corollary 3.11. By means of the hodograph (reciprocal) transformation $((y, s) \to (x, t))$

$$x=y_0+\int_{y_0}^{y}\rho\left(y',s'\right)dy',\quad t=s,$$

the equation CSP is obtained from CCD equation according to Bishop frame [25].

References

[1] K. Konno and H. Oono, *New coupled dispersionless equations*, J. Phys. Soc. Jpn. **63** (1994) 377–378.

[2] K. Konno and H. Kakuhata, *Interaction among growing, decaying and stationary solitons for coupled and stationary solitons*, J. Phys. Soc. Jpn. **64** (1995) 2707–2709.

[3] H. Kakuhata and H. Konno, *A generalization of coupled integrable, dispersionless system*, J. Phys. Soc. Jpn. **65** (1996) 340–341.

[4] H. Kakuhata and K. Konno, *Lagrangian, Hamiltonian and conserved quantities for coupled integrable, dispersionless equations*, J. Phys. Soc. Jpn. **65** (1996) 1–2.

[5] T. Schäfer and C. E. Wayne, *Propagation of ultra-short optical pulses in cubic nonlinear media*, Physica D. **196** (2004) 90–105.

[6] B. F. Feng, *Complex short pulse and coupled complex short pulse equations*, Physica D. **297** (2015) 62–75.

[7] Y. Chung, C.K.R.T. Jones, T. Schäfer and C.E. Wayne, *Ultra-short pulses in linear and nonlinear media*, Nonlinearity **18** (2005) 1351–1374.

[8] A. Sakovich and S. Sakovich, *The short pulse equation is integrable*, J. Phys. Soc. Jpn. **74** (2005) 239–241.

[9] J.C. Brunelli, *The short pulse hierarchy*, J. Math. Phys. **46** (2005) 123–507.

[10] J.C. Brunelli, *The bi-Hamiltonian structure of the short pulse equation*, Phys. Lett. A **353** (2006) 475–478.

[11] Y. Matsuno, *Multisoliton and multibreather solutions of the short pulse model equation*, J. Phys. Soc. Jpn. **76** (2007) 08–4003.

[12] Y. Matsuno, *Periodic solutions of the short pulse model equation*, J. Math. Phys. **49** (2008) 073508.

[13] A. Yariv and P. Yeh, *Optical Waves in Crystals: Propagation and Control of Laser Radiation*, Wiley New York (1984).

[14] B.F. Feng, K. Maruno and Y. Ohta, *Self-adaptive moving mesh schemes for short pulse type equations and their Lax pairs*, Pac. J. Math. Ind **6** (2014) 7–20.

[15] S.Y. Lou and G.F. Yu, *A generalization of the coupled integrable dispersionless equations*, Mathematical Methods in the Applied Sciences **39** (2016) 4025–4034.

[16] Y.H. Wang, *On the integrability of the (1+1)–dimensional and (2+1)–dimensional Ito equations*, Mathematical Methods in the Applied Sciences **38** (2014) 138–144.

[17] S.F. Shen, B.F. Feng and Y. Ohta, *From the real and complex coupled dispersionless equations to the real and complex short pulse equations*, Stud. Appl. Math. **136** (2016) 64–88.

[18] L.R. Bishop, *There is more than one way to frame a curve*, Amer. Math. Monthly **82** (1975) 246–251.

[19] A. J. Hanson and H. Ma, *Parallel Transport Approach to Curve Framing*, Indiana University (1995).

[20] T. Körpınar, V. Asil and S. Baş, *On characterization inextensible flows of curves according to Bishop frame in* E^3, Revista Notas de Matematica **7** (2011) 37–45.

[21] S. Yılmaz and M. Yılmaz, *A new version of Bishop frame and application to spherical images*, J. Math. Anal.Appl. **371** (2010) 764–776.

[22] E. Özyılmaz, *Classical differential geometry of curves according to type-2 Bishop trihedra*, Math. Comput. Appl. **16** (2011) 858–867.

[23] K. Konno and H. Kakuhata, *Novel solitonic evolutions in a coupled integrable, dispersionless system*, J. Phys. Soc. Jpn. **65** (1996) 713–721.

[24] C. Rogers and W.K. Schief, *Bäcklund and Darboux Transformations: Geometry and Modern Applications in Soliton Theory*, Cambridge University Press (1996).

[25] B.F. Feng, J. Inoguchi, K. Kajiwara, K. Maruno and Y. Ohta , *Discrete integrable systems and hodograph transformations arising from motions of discrete plane curves*, J. Phys. A: Math. Theor. **44** (2011) 395201.

DOI 10.2478/9788395793882-005

H-bases and lifting problem for homogeneous ideals

Erol Yılmaz [1] and Sibel Cansu[2]

[1,2] Department of Mathematics, Bolu Abant Izzet Baysal University, Bolu, Turkey
E-mail: yilmaz_e2@ibu.edu.tr[1], kilicarslan_s@ibu.edu.tr[2]

Abstract

The lifting problem for homogeneous ideals in a polynomial ring over a field is studied. A division algorithm for H-bases is given. Using this division algorithm, a new method for finding the liftings of a homogeneous ideal is developed. This method was compared with the current methods. The results are demonstrated with examples.

2010 Mathematics Subject Classification. **13P10** 14A03
Keywords. H-bases, lifting problem, division algorithm, syzygy module.

1 Introduction

Let $A = K[x_1, x_2, \ldots, x_{n-1}]$ and $B = K[x_1, x_2, \ldots, x_{n-1}, x_n]$ where K is a field.

Definition 1.1. Let J be a homogeneous ideal in the polynomial ring A. A homogeneous ideal I in B is called lifting of J if

(i) x_n is not a zero divisor on B/I; and

(ii) $J = \langle f(x_1, x_2, \ldots, x_{n-1}, 0) : f \in I \rangle$.

The problem of finding the liftings of a given homogeneous ideal first suggested in [1] and fully explained in [2]. Then this interesting problem has been investigated by many researchers (see [3, 4, 5, 6]).

More recently a new computational method for finding the liftings of a given homogeneous ideal is given [7]. This method involves some Gröbner basis computations. Here we will give only a few definitions that are sufficient to explain the method given in [7]. We refer to [8] for a detailed treatment of Gröbner bases.

For a given a monomial order $<$ and a polynomial f in a polynomial ring, $\mathrm{LT}(f)$ denotes the leading term of f with respect to $<$. If J is an ideal in this polynomial ring $\mathrm{LT}(J) = \langle \mathrm{LT}(f) : f \in J \rangle$ and $\mathcal{N}_J$ is the set of monomials which are not in $\mathrm{LT}(J)$.

Now we are ready to explain the method given in [7]. Starting with a homogeneous ideal $J = \langle f_1, \ldots, f_s \rangle \subset A$ compute a Gröbner basis $\{h_1, \ldots, h_t\}$ of J with respect to a given monomial order $<$ in A. After that define the polynomials

$$g_i = h_i + \sum_{\substack{x^\alpha x_n \in \mathcal{N}_J \\ \deg(x^\alpha x_n) = \deg(h_i)}} C_{i\alpha} x^\alpha x_n$$

where $x^\alpha = x_1^{\alpha_1} x_2^{\alpha_2} \ldots x_n^{\alpha_n}$.

The set of polynomials $G = \{g_1, \ldots, g_t\}$ is a Gröbner basis for the ideal $I = \langle G \rangle$ with respect to a special extension of monomial order$<$ into a monomial order in B if and only if I is a lifting

Tbilisi Mathematical Journal Special Issue (IECMSA-2019), pp. 49–58.
Tbilisi Centre for Mathematical Sciences.

Received by the editors: 07 December 2019.
Accepted for publication: 03 January 2020.

of J. Buchberger's criterion for Gröbner bases implies the remainders of S-polynomials $S(g_i, g_j)$ on division by G is zero for $1 \leq i < j \leq t$. Hence equalizing the coefficients of remainders to zero for each division gives us the conditions that the parameters must satisfy for I to be a lifting of J. Theoretical background of this method depends on a reformulation of a theorem given [4] in terms of Gröbner bases. In the original theorem however the lifting problem is related to the H-bases not to the Gröbner bases. Therefore, this method is against to the nature of the problem. It finds a Gröbner basis of a lifting which is not essential. Because of this, the method involves many unnecessary computations.

The first attempt to use H-bases to solve the lifting problem was given in [6]. The authors gave a criterion for H-bases in terms of syzygy modules which is very similar to Buchbergers criterion for Gröbner bases. Because they could not use the division algorithm on H-bases, they were not able to give a complete method for solving the lifting problem.

In this paper, we use the vector space character of homogeneous ideals to define a kind of division on homogeneous polynomials. Combining the ideas given in [6] with this division algorithm, we developed a new method for solving the lifting problem for homogeneous ideals. This method is similar to the method given in [7] but it uses H-bases instead of Gröbner bases. Finally, we compare these two methods and explain the advantages of our method.

2 H-bases

If $f = f_d + f_{d-1} + \cdots + f_1 + f_0$ where each $f_i \in B$ is a homogeneous polynomial of degree i and $f_d \neq 0$, then $H(f) = f_d$ is called the leading form of f. For an ideal I of A, $H(I) = \langle H(f) : f \in I\rangle$.

Definition 2.1. A set of polynomials $\{f_1, f_2, \ldots, f_s\} \subset B$ is an H-basis for the ideal $I = \langle f_1, f_2, \ldots, f_s\rangle$ provided that $H(I) = \langle H(f_1), H(f_2), \ldots, H(f_s)\rangle$.

Let $f = f_d + f_{d-1} + \cdots + f_1 + f_0$ be a polynomial in the polynomial ring A. The homogenization of f with respect to x_n is defined to be the polynomial $f^h = f_d + x_n f_{d-1} + \cdots + x_n^{d-1} f_1 + x_n^d f_0 \in B$. For an ideal I of A the homogenization of I, denoted by I^h, is the ideal $I^h = \langle f^h : f \in I\rangle \subseteq B$.

Let f be a homogeneous polynomial in B, then the dehomogenization of f is the polynomial $f_a = f(x_1, x_2, \ldots, x_{n-1}, 1) \in A$. For a homogeneous ideal I of B, the dehomogenization of I is the ideal $I_a = \langle f_a : f \in I\rangle \subseteq A$.

Lemma 2.2. [4, Lemma 2.4] Let I be a homogeneous ideal of B. Then the following conditions are equivalent.

(i) x_n is not a zero divisor on B/I,

(ii) $I = (I_a)^h$,

(iii) $H(I_a) = I(x_1, x_2, \ldots, x_{n-1}, 0)$.

The next theorem gives a relation between the lifting problem and H-bases. A Gröbner basis version of this theorem is used in [7].

Theorem 2.3. [4, Theorem 2.5] Let $J = \langle f_1, f_2, \ldots, f_s\rangle$ be a homogeneous ideal of A.

(i) Let $g_i = f_i + R_i$ with $deg(R_i) < deg(f_i)$ for $1 \leq i \leq s$ and $I = \langle g_1, g_2, \ldots, g_s\rangle \subseteq A$. If $\{g_1, g_2, \ldots, g_s\}$ is an H-basis for I, then $I^h = \langle g_1^h, g_2^h, \ldots, g_s^h\rangle$ is a lifting of J. Conversely,

(ii) If I is a lifting of J, then there exist polynomials $R_1, R_2, \ldots, R_s \in A$ such that $deg(R_i) < deg(f_i)$ for every i, $\{g_1, g_2, \ldots, g_s\}$ is an H-basis and $I = \langle g_1^h, g_2^h, \ldots, g_s^h \rangle$.

This theorem is used in original form to find the liftings of a homogeneous ideal in[6]. We try to explain their method but we need to define the syzygy module of a set of polynomials.

Definition 2.4. For an s-tuple of polynomials $(f_1, \ldots, f_s)$, the module generated by following set of s-tuple of polynomials

$$\{(h_1, \ldots, h_s) : h_1 f_1 + \cdots + h_s f_s = 0\}$$

is called syzygy module of $(f_1, \ldots, f_s)$ and denoted by $syz(f_1, \ldots, f_s)$.

The following theorem gives a criterion for a set of polynomials to be an H-basis. This criterion is H-basis version of Buchberger's criterion for a Gröbner basis.

Theorem 2.5. [6, Theorem 2.4] Let $I = \langle h_1, h_2, \ldots, h_t \rangle \subseteq K[x_1, \ldots, x_n]$. Let the columns of the $t \times l$ matrix $S = (s_{ij})$ be a generating set of $syz(H(h_1), \ldots, H(h_t))$. We may assume further that each $s_{ji} f_j$ is a homogeneous polynomial of same degree for $j = 1, \ldots, t$. Then $\mathcal{H} = \{h_1, h_2, \ldots, h_t\}$ is an H-basis for I if and only if

$$q_i = \sum_{j=1}^{t} s_{ji} h_j = \sum_{j=1}^{t} a_{ji} h_j, \quad 1 \leq i \leq l$$

for some $a_{ji} \in K[x_1, \ldots, x_n]$ such that $deg(q_i) = \max\{deg(a_{ji} h_j), j = 1, \ldots, t\}$.

The following method for finding the liftings of a homogeneous ideal is proposed in [6]: Given an ideal $J = \langle f_1, \ldots, f_r \rangle \subseteq K[x_1, \ldots, x_n]$ where f_i's are homogeneous, define

$$g_i = \sum_{deg(x^\gamma) < deg(f_i)} C_{i\gamma} x^\gamma, \quad h_i = f_i + g_i.$$

Furthermore, for each q_i in Theorem 2.5 define

$$a_{ji} = \sum_{deg(x^\gamma) < deg(q_i) - deg(h_j)} D_{i\gamma} x^\gamma.$$

Then compare the coefficient of monomials of the equation given Theorem 2.5 to find relations among the parameters $C_{i\gamma}$'s and $D_{ij\gamma}$'s. Then $I = \langle h_1, h_2, \ldots, h_r \rangle$ is an element of the liftings of J if and only if the coefficients of h_i's satisfy these relations. This is not a convenient method because there are extra parameters $D_{ij\gamma}$'s. Even though in their example they are able to solve these extra parameters in terms of $C_{i\gamma}$'s, there is no guarantee that this will always occur.

The effectiveness of the method proposed in [7] comes from the usage of the division algorithm. This is the weak part of the method in [6]. Because of this we define a division process for H-bases.

3 A division algorithm for H-bases

Let P_d be the vector space of homogeneous polynomials of degree d in variables $x_1, x_2, \ldots, x_n$ over a field K. For brevity we denote the monomial $x_1^{\alpha_1} x_2^{\alpha_2} \ldots x_{n-1}^{\alpha_{n-1}} x_n^{\alpha_n}$ by x^α. It is well-known that

$$\mathcal{B} = \{x^\alpha : \deg(x^\alpha) = \alpha_1 + \alpha_2 + \ldots + \alpha_n = d\}$$

is a basis for P_d.

Let $I = \langle f_1, f_2, \ldots, f_s \rangle \subseteq K[x_1, x_2, \ldots, x_n]$ where f_i's are homogeneous polynomials. $V_d(I) = \{f \in I : deg(f) = d \quad or \quad f = 0\}$ is a subspace of P_d. Furthermore, the polynomials of the form $x_1^{\alpha_1} \ldots x_n^{\alpha_n} f_i$ where $\deg(x^\alpha) + \deg(f_i) = d$ spans $V_d(I)$.

For a given homogeneous polynomial $f \in K[x_1, x_2, \ldots, x_n]$ of degree d, $f \in I$ if and only if $f \in V_d(I)$. We can decide whether $f \in V_d(I)$ or not as follows:

Let $f = \sum_{\deg(x^\alpha)=d} a_i x^\alpha$ where $a_i \in K$. We have to find c_{β_j}'s in K such that

$$f = \sum_{\deg(x^\beta)+\deg(f_j)=d} c_{\beta_j} x^\beta f_j.$$

This is just a system of linear equations and can be solved linear algebra techniques. Construct the matrix M whose columns are coordinate vectors of $x^\beta f_j$'s with respect to $\mathcal{B}$. Also the last column of M is the coordinate vector of f with respect to $\mathcal{B}$. If M' is the row reduced echelon form of M, then these matrices will give the solution of same system since they are row equivalent. If M' has any row like $(0, 0, \ldots, 0, L(a_1, a_2, \ldots))$ where L is a linear function, then the system has no solution unless $L(a_1, a_2, \ldots) = 0$. Such rows will give relations between a_i's for f to be in $V_d(I)$.

Example 3.1. Consider the homogeneous ideal

$$I = \langle f_1, f_2, f_3 \rangle = \langle x_1^2 + x_1 x_3, x_1 x_2 + x_2 x_3, x_1^3 + x_1 x_2^2 + x_2^2 x_3 \rangle$$

and the polynomial

$$g = a_1 x_1^3 + a_2 x_1^2 x_2 + a_4 x_1 x_2^2 + a_7 x_2^3 + a_3 x_1^2 x_3 + a_5 x_1 x_2 x_3 + a_8 x_2^2 x_3 + a_6 x_1 x_3^2 + a_9 x_2 x_3^2 + a_{10} x_3^3$$

of degree 3. We try to obtain the equations that a_i's must satisfy for the polynomial g to be in $V_3(I)$.

Clearly, $\{x_1 f_1, x_2 f_1, x_3 f_1, x_1 f_2, x_2 f_2, x_3 f_2, f_3\}$ is a spanning set for the vector space $V_3(I)$. Consider the augmented matrix M,

	$x_1 f_1$	$x_2 f_1$	$x_3 f_1$	$x_1 f_2$	$x_2 f_2$	$x_3 f_2$	f_3	g
x_1^3	1	0	0	0	0	0	1	a_1
$x_1^2 x_2$	0	1	0	1	0	0	0	a_2
$x_1^2 x_3$	1	0	1	0	0	0	0	a_3
$x_1 x_2^2$	0	0	0	0	1	0	1	a_4
$x_1 x_2 x_3$	0	1	0	1	0	1	0	a_5
$x_1 x_3^2$	0	0	1	0	0	0	0	a_6
x_2^3	0	0	0	0	0	0	0	a_7
$x_2^2 x_3$	0	0	0	0	1	0	1	a_8
$x_2 x_3^2$	0	0	0	0	0	1	0	a_9
x_2^3	0	0	0	0	0	0	0	a_{10}

After applying a series of elementary row operations, the row reduced echelon form of M, say $M^{'}$, will be

	x_1f_1	x_2f_1	x_3f_1	x_1f_2	x_2f_2	x_3f_2	f_3	g
x_1^3	1	0	0	0	0	0	0	$-a_6+a_3$
$x_1^2x_2$	0	1	0	1	0	0	0	a_2
$x_1^2x_3$	0	0	1	0	0	0	0	a_6
$x_1x_2^2$	0	0	0	0	1	0	0	$a_4-a_6+a_3-a_1$
$x_1x_2x_3$	0	0	0	0	0	1	0	a_5-a_2
$x_1x_3^2$	0	0	0	0	0	0	1	$a_6-a_3+a_1$
x_2^3	0	0	0	0	0	0	0	a_7
$x_2^2x_3$	0	0	0	0	0	0	0	a_8-a_4
$x_2x_3^2$	0	0	0	0	0	0	0	$a_9-a_5+a_2$
x_2^3	0	0	0	0	0	0	0	a_{10}

Hence, from the last 4 rows of $M^{'}$, $g \in V_3(I)$ if and only if

$$\begin{aligned} a_7 &= 0 \\ a_4 &= a_8 \\ a_9 &= a_5 - a_2 \\ a_{10} &= 0 \end{aligned}$$

Under these conditions,
$g = [(-a_6+a_3)x_1 + a_2x_2 + a_6x_3]f_1 + [(a_4-a_6+a_3-a_1)x_2 + (a_5-a_2)x_3]f_2 + (a_6-a_3+a_1)f_3.$

Now we are ready to define a new division process.

Definition 3.2. Let $\mathcal{H} = \{f_1, f_2, \ldots, f_s\} \subseteq K[x_1, \ldots, x_n]$. For $f \in K[x_1, \ldots, x_n]$ we say f reduces to $\widetilde{f}$ modulo $\mathcal{H}$, written
$f \longrightarrow_{\mathcal{H}} \widetilde{f}$, if

$$\widetilde{f} = f - (a_1f_1 + \cdots + a_sf_s)$$

for some homogeneous polynomials $a_1, \ldots, a_s$ satisfying

$$H(f) = a_1H(f_1) + \cdots + a_sH(f_s)$$

and $deg(a_i) = deg(f) - deg(f_i)$.

We say f completely reduce to r, written $f \longrightarrow_{\mathcal{H}}^{+} r$, if there exists a sequence of polynomials $g_1, \ldots, g_t$ such that

$$f \rightarrow_{\mathcal{H}} g_1 \rightarrow_{\mathcal{H}} g_2 \rightarrow_{\mathcal{H}} \cdots \rightarrow_{\mathcal{H}} g_t \rightarrow_{\mathcal{H}} r$$

and no homogeneous part of r is in $\langle H(f_1), \ldots, H(f_s) \rangle$.

Lemma 3.3. Let $\mathcal{H} = \{f_1, f_2, \ldots, f_s\} \subseteq K[x_1, \ldots, x_n]$. For every $f \in K[x_1, \ldots, x_n]$, there exists $r \in K[x_1, \ldots, x_n]$ such that $f \longrightarrow_{\mathcal{H}}^{+} r$.

Proof. The desired polynomial r can be found using the following algorithm.

$h := f, r := 0$

WHILE $h \neq 0$ DO

IF $H(h) \in \langle H(f_1), \ldots, H(f_s) \rangle$ THEN

$h := \widetilde{h}$ where $h \longrightarrow_{\mathcal{H}} \widetilde{h}$

ELSE

$h := h - H(h)$

$r := r + H(h)$

Q.E.D.

Let us illustrate the algorithm with an example.

Example 3.4. Let $\mathcal{H} = \{f_1, f_2\} = \{x_1^3 + x_2x_3, x_1x_2 + x_3\}$ and $f = x_1^3 + x_1x_2^2 + x_2x_3 + x_1^2 + x_3$. We will apply the algorithm to find the polynomial r.

- Here $h = f$ and $r = 0$. Using technique given in Example 3.1,
$$H(h) = x_1^3 + x_1x_2^2 = H(f_1) + x_2H(f_2).$$
So, $h = \widetilde{h} = h - f_1 - x_2f_2 = x_1^2 - x_2x_3 + x_3$.
- Since $H(h) = x_1^2 - x_2x_3 \notin \langle H(f_1), H(f_2) \rangle$,
$$h = h - H(h) = x_3$$
and
$$r = 0 + H(h) = x_1^2 - x_2x_3.$$
- Now, $h \neq 0$ and $H(h) = x_3 \notin \langle H(f_1), H(f_2) \rangle$. So,
$$h = h - H(h) = 0$$
and
$$r = r + H(h) = x_1^2 - x_2x_3 + x_3.$$

Since $h = 0$, the algorithm ends with $r = x_1^2 - x_2x_3 + x_3$. Therefore, $f \longrightarrow_{\mathcal{H}}^{+} x_1^2 - x_2x_3 + x_3$.

4 New method

The following results give some relations between H-bases and the division algorithm defined on the previous section.

Lemma 4.1. Let $\mathcal{H} = \{f_1, f_2, \dots, f_s\} \subseteq K[x_1, \dots, x_n]$ and $I = \langle \mathcal{H} \rangle$. $\mathcal{H}$ is an H-basis for I if and only if for every $f \in I$, $f \longrightarrow^+_{\mathcal{H}} 0$.

Proof. Suppose that $\mathcal{H} = \{f_1, f_2, \dots, f_s\}$ is an H-basis for $I = \langle \mathcal{H} \rangle$ and $f \in I$. Then there exist homogeneous polynomials $a_1, \dots, a_s \in K[x_1, \dots, x_n]$ satisfying $deg(a_i) = deg(f) - deg(f_i)$ such that $H(f) = a_1 H(f_1) + \cdots + a_s H(f_s)$. Notice that these polynomials can be obtained by the simple linear algebra techniques. Now define $\widetilde{f} = f - (a_1 f_1 + \cdots + a_s f_s)$. So, $f \longrightarrow_{\mathcal{H}} \widetilde{f}$. It is clear that $deg(\widetilde{f}) < deg(f)$ and $\widetilde{f} \in I$. Hence we can apply same process to the polynomial $\widetilde{f}$ and continue doing this until we reach to zero polynomial.

Conversely, suppose that $f \in I$ and $f \longrightarrow^+_{\mathcal{H}} 0$. Then there exists a sequence of polynomials $g_1, \dots, g_t$ such that

$$f \rightarrow_{\mathcal{H}} g_1 \rightarrow_{\mathcal{H}} g_2 \rightarrow_{\mathcal{H}} \cdots \rightarrow_{\mathcal{H}} g_t \rightarrow_{\mathcal{H}} 0.$$

Since $f \rightarrow_{\mathcal{H}} g_1$, there exist homogeneous polynomials $b_1, \dots, b_s$ such that $H(f) = b_1 H(f_1) + \cdots + b_s H(f_s)$ which implies that $H(f) \in \langle H(f_1), \dots, H(f_s) \rangle$. Hence $H(I) = \langle H(f_1), \dots, H(f_s) \rangle$, that means $\mathcal{H} = \{f_1, f_2, \dots, f_s\}$ is an H-basis.

Q.E.D.

Lemma 4.2. Let $\mathcal{H} = \{f_1, f_2, \dots, f_s\} \subseteq K[x_1, \dots, x_n]$ and $I = \langle \mathcal{H} \rangle$. For every $f \in I$, $f \longrightarrow^+_{\mathcal{H}} 0$ if and only if there exist polynomials $a_1, \dots, a_s \in K[x_1, \dots, x_n]$ such that $f = a_1 f_1 + \cdots + a_s f_s$ and $deg(f) = \max_{1 \le i \le s} \{deg(a_i f_i)\}$.

Proof. Suppose that for every $f \in I$ there exist polynomials $a_1, \dots, a_s \in K[x_1, \dots, x_n]$ such that $f = a_1 f_1 + \cdots + a_s f_s$ and $deg(f) = \max_{1 \le i \le s} \{deg(a_i f_i)\}$. Then

$$H(f) = \sum_{deg(f) = deg(a_i f_i)} H(a_i) H(f_i).$$

$H(I) = \langle H(f_1), \dots, H(f_s) \rangle$ means that $\mathcal{H} = \{f_1, f_2, \dots, f_s\}$ is an H-basis. Hence $f \longrightarrow^+_{\mathcal{H}} 0$ by above lemma.

Suppose that $f \in I$ and $f \longrightarrow^+_{\mathcal{H}} 0$. Then

$$f = g_0 \rightarrow_{\mathcal{H}} g_1 \rightarrow_{\mathcal{H}} g_2 \rightarrow_{\mathcal{H}} \cdots \rightarrow_{\mathcal{H}} g_t = 0$$

for some polynomials $g_1, \dots, g_t \in K[x_1, \dots, x_n]$. If

$$H(g_{i-1}) = \sum_{j=1}^{s} a_{ij} H(f_j)$$

for $i = 1, \dots, t$, then

$$f = \sum_{j=1}^{s} \sum_{i=1}^{t} a_{ij} f_j.$$

Q.E.D.

Using above lemmas, we can rewrite Theorem 2.5 with the new notation.

Theorem 4.3. Let $I = \langle h_1, h_2, \ldots, h_t \rangle \subseteq K[x_1, \ldots, x_n]$. Let the columns of the $t \times l$ matrix $S = (s_{ij})$ be a generating set of $syz(H(h_1), \ldots, H(h_t))$. We may assume further that each $s_{ji}f_j$ is a homogeneous polynomial of same degree for $j = 1, \ldots, t$. Then $\mathcal{H} = \{h_1, \ldots, h_t\}$ is an H-basis for I if and only if

$$q_i = \sum_{j=1}^{t} s_{ji}h_j \longrightarrow_{\mathcal{H}}^{+} 0, \quad 1 \leq i \leq l.$$

Now we are ready to explain the new method for findings of the liftings of a given homogeneous ideal $J = \langle f_1, \ldots, f_s \rangle \subset A$. First of all, we need to find a generating set for syz$(f_1, , f_s)$. Secondly, we define the polynomials

$$g_i = f_i + \sum_{\deg(x^\gamma) < \operatorname{def}(f_i} C_{i\gamma}x^\gamma \quad 1 \leq i \leq s.$$

Then for each syzygy $(t_1, \ldots, t_s)$ in the generating set of syz$(f_1, , f_s)$ define the polynomial $q_t = t_1g_1 + \cdots + t_sg_s$. Theorem 4.3 implies $(H) = \{g_1, \ldots, g_s\}$ is an H-basis, in other words $I = \langle (H) \rangle$ is a lifting for J, if and only if every $q_t \longrightarrow_{\mathcal{H}}^{+} 0$. Therefore when applying the division algorithm, described in Lemma 3.3, to each q_t the conditions that the parameters $C_{i\gamma}$'s must satisfy for I to be a lifting of J can be obtained.

Let us illustrate the method with an example.

Example 4.4. Condider the ideal $J = \langle f_1, f_2, f_3 \rangle = \langle x_1^2, x_1x_2, x_2^4 + x_1x_3^3 \rangle$. Define the polynomials:

$g_1 = f_1 + C_1x_1 + C_2x_2 + C_3x_3 + C_4,$

$g_2 = f_2 + C_5x_1 + C_6x_2 + C_7x_4 + C_8$ and

$g_3 = f_3 + C_9x_1^3 + C_{10}x_1^2x_2 + C_{11}x_1^2x_3 + C_{12}x_1x_2^2 + C_{13}x_1x_2x_3 + C_{14}x_1x_3^2 + C_{15}x_2^3 + C_{16}x_2^2x_3 + C_{17}x_2x_3^2 + C_{18}x_3^3 + C_{19}x_1^2 + C_{20}x_1x_2 + C_{21}x_1x_3 + C_{22}x_2^2 + C_{23}x_2x_3 + C_{24}x_3^2 + C_{25}x_1 + C_{26}x_2 + C_{27}x_3 + C_{28}.$

The syzygy module syz(f_1, f_2, f_3) can be generated $t_1 = (x_2, -x_1, 0), t_2 = (0, f_3, -f_2)$ and $t_3 = (x_3^3, x_2^3, -x_1)$. The details of computation of generators of syzygy modules can be found in [9]. Then we define polynomials:

$q_1 = x_2g_1 - x_1g_2 = -C_5x_1^2 + (C_1 - C_6)x_1x_2 + C_2x_2^2 - C_7x_1x_3 + C_3x_2x_3 - C_8x_1 + C_4x_2, q_2 = f_3g_2 - f_2g_3$ and $q_3 = x_3^3g_1 + x_23g_2 - x_1g_3$.

The ideal $I = \langle g_1, g_2, g_3 \rangle$ is a lifting of J if and only if $q_1 \longrightarrow_{\mathcal{H}}^{+} 0$ for $i = 1, 2, 3$. Let us apply division algorithm to q_1.

$$H(q_1) = -C_5x_1^2 + (C_1 - C_6)x_1x_2 + C_2x_2^2 - C_7x_1x_3 + C_3x_2x_3.$$

Using the technique given in Example 3.1, $H(q_1) = -C_5f_1 + (C_1 - C_6)f_2$ under the condition $C_2 = C_3 = C_7 = 0$.

Let

$$\tilde{q_1} = q_1 + C_5g_1 - (C_1 - C_6)g_2 = (C_5C_6 - C_8)x_1 + (C_4 - C_1C_6 + C_6^2)x_2 + C_4C_5 - C_1C_8 + C_6C_8.$$

This a polynomial of degree 1, so it goes to the remainder. Since the remainder should be zero, the equations $C_8 = C_5C_6$ and $C_4 = C_1C_6 - C_6^2$ are obtained.

Applying same process to q_2 produce the following equations:

$C_6 = C_1 - C_{18}$,

$C_{24} = C_5C_{17} + C_{14}C_{18}$,

$C_{27} = -C_5^2C_{16} - C_5C_{13}C_{18} - C_{11}C_{18}^2 + C_{18}C_{21} + C_5C_{23}$ and

$C_{28} = -C_5^4 + C_5^3C_{15} + C_5^2C_{12}C_{18} + C_5C_{10}C_{18}^2 + C_9C_{18}^3 - C_{18}^2C_{19} - C_5C_{18}C_{20} - C_5^2C_{22} + C_{18}C_{25} + C_5C_{26}$

Replacing $C_6 = C_1 - C_{18}$ into equations $C_8 = C_5C_6$ and $C_4 = C_1C_6 - C_6^2$, we also get $C_4 = C_1C_{18} - C_{18}^2$ and $C_8 = C_1C_5 - C_5C_{18}$.

The division of q_3 does not produce additional equations. Hence $I = \langle g_1, g_2, g_3\rangle$ is a lifting of J if

$g_1 = x_1^2 + C_1x_1 + C_1C_{18} - C_{18}^2$,

$g_2 = x_1x_2 + C_5x_1 + (C_1 - C_{18})\,x_2 + C_1C_5 - C_5C_{18}$ and

$g_3 = x_2^4 + x_1x_3^3 + C_9x_1^3 + C_{11}x_1^2x_3 + C_{23}x_2x_3 + C_{13}x_1x_2x_3 + C_{16}x_2^2x_3 + (C_5C_{17} + C_{14}C_{18})\,x_3^2 + C_{14}x_1x_3^2 + C_{17}x_2x_3^2 + C_{18}x_3^3 + C_{19}x_1^2 + C_{10}x_1^2x_2 + C_{22}x_2^2 + C_{12}x_1x_2^2 + C_{15}x_2^3 + C_{25}x_1 + C_{26}x_2 + \left(-C_5^2C_{16} - C_5C_{13}C_{18} - C_{11}C_{18}^2 + C_{18}C_{21} + C_5C_{23}\right)x_3 - C_5^4 + C_5^3C_{15} + C_5^2C_{12}C_{18} + C_5C_{10}C_{18}^2 + C_9C_{18}^3 - C_{18}^2C_{19} - C_5C_{18}C_{20} - C_5^2C_{22} + C_{18}C_{25} + C_5C_{26}$.

Starting with a generating set for the homogeneous ideal J, the one need to add new polynomials to the generating set unless the original set is a Gröbner basis in the method suggested in [7]. This is the most important handicap of that method. In our method however we always use the given generating set of the ideal J. On the other hand, our method requires a generating set for the syzygy module. The best-known method for computation of a syzygy module uses Gröbner bases (see [9]). Even if the Gröbner basis of J is computed for finding a syzygy module, many S-polynomials do not produce a syzygy for J at the end. Hence the number of syzygies to be considered in our method is generally much less than the number of S-polynomials to be considered in the method given in[7]. Furthermore, we may only need a Gröbner basis for J not for I which contains polynomials with many parameters. The Gröbner basis computation with parameters might be very complicated.

References

[1] A. V. Geramita, D. Gregory and L. Roberts, *Monomial ideals and points in projective space*, J. Pure Appl. Algebra **40 (1)** (1986) 33–62.

[2] M. Roitman, *On the lifting problem for homogeneous ideals in polynomial rings*, J. Pure Appl. Algebra **51(1-2)** (1988) 205–215.

[3] Leslie G. Roberts, *On the lifting problem over an algebraically closed field*, C. R. Math. Rep. Acad. Sci. Canada **11(1)** (1989) 35–38.

[4] G. Carrá Ferro and , L. Robbiano, *On super G-bases*, J. Pure Appl. Algebra **68 (3)** (1990) 279–292.

[5] J. Migliore and U. Nagel, *Lifting monomial ideals*, Commun. Algebra **28(12)** (2000) 5679–5701.

[6] T. Luo and E. Yılmaz, *On the lifting problem for homogeneous ideals*, J. Pure Appl. Algebra **162(2-3)** (2001) 327–335.

[7] C. Bertone, F. Cioffi, M. Guida and M. Roggero , *The scheme of liftings and applications*, J. Pure Appl. Algebra **220** 2016 34–54.

[8] D. Cox, J. Little and D. O'Shea, *Ideals, Varieties and Algorithms*, Springer (2015).

[9] D. Cox, J. Little and D. O'Shea, *Using Algebraic Geometry*, Springer (2005).

DOI 10.2478/9788395793882-00

On generalization of dual Fibonacci octonions

Serpil Halici and Adnan Karataş*

Pamukkale University, Faculty of Arts and Sciences, Department of Mathematics, Denizli/TURKEY
*Corresponding Author
E-mail: shalici@pau.edu.tr, adnank@pau.edu.tr

Abstract

In this study, we examine all the second order linear recurrence relations over dual octonions. Hence, we generalize Fibonacci-like relations over dual octonions. For this purpose we use the well-known Horadam sequence and obtain some fundamental and new identities involving elements of this generalized sequence.

2010 Mathematics Subject Classification. **11B39** 17A20
Keywords. Fibonacci numbers and generalization, Horadam sequence, octonions.

1 Introduction

A dual octonion can be defined as $\hat{K} = (k,l) = k + l\varepsilon$ where $k,l \in \mathbb{O}$. Let $\mathbb{D}$ be the set of dual octonions;

$$\mathbb{D} = \{\hat{K} | \hat{K} = (k,l) = k + l\varepsilon; \quad k,l \in \mathbb{O} \text{ with } \varepsilon^2 = 0\}. \tag{1.1}$$

The addition on the set $\mathbb{D}$ is given by

$$\hat{K} + \hat{L} = (k,l) + (m,n) = (k+m, l+n) = (k+m) + (l+n)\varepsilon. \tag{1.2}$$

It should be noted that there are different multiplication tables for octonions which can be obtained by Cayley-Dickson process [4, 5]. We define the multiplication of the bases elements as $e_i e_j = e_k$. The set of indices $\{i,j,k\}$ being one of $\{1,2,3\}$, $\{1,4,5\}$, $\{1,7,6\}$, $\{2,4,6\}$, $\{2,5,7\}$, $\{3,4,7\}$ and $\{3,6,5\}$. Note that if a different multiplication is chosen, then one should make all calculations accordingly (see, [1,4]). The multiplication of two dual octonions can be defined as

$$\widehat{K}\widehat{L} = (km, kn + lm). \tag{1.3}$$

The conjugate of a dual octonion $\widehat{K}$ is going to be denoted by $\overline{\widehat{K}}$ and this will be written as follows.

$$\overline{\widehat{K}} = (\overline{k}, \overline{l}) = \overline{k} + \overline{l}\varepsilon. \tag{1.4}$$

Norm is defined by the help of conjugate as follows;

$$N(\widehat{K}) = (k,l)(\overline{k},\overline{l}) = (\overline{k},\overline{l})(k,l) = (k\overline{k}, k\overline{l} + l\overline{k}). \tag{1.5}$$

We also note that the definition of a dual unit ε causes the norm to be degenerate which means that the norm can be equal to zero for some non zero dual octonions. To explain this situation let us calculate the norm for $\widehat{K} = 1\varepsilon$;

$$N(\widehat{K}) = 1\varepsilon\overline{1}\varepsilon = 1\varepsilon^2 = 0. \tag{1.6}$$

Tbilisi Mathematical Journal Special Issue (IECMSA-2019), pp. 59–67.
Tbilisi Centre for Mathematical Sciences.

Received by the editors: 08 October 2019.
Accepted for publication: 02 January 2020.

For detailed knowledge, we refer the interested readers to [1] and [3].

The rest of this paper is organized as follows. Section 2 reviews basic properties of Horadam sequence which will be used in the following sections. In Section 3, the dual Horadam octonions are introduced and investigated. Moreover, some important identities and equalities involving these octonions are obtained. And finally, in Section 4, the generalization is made in this study is summarized.

2 The Horadam sequence

In this section, we focus on the second order recursive relations over the dual octonions. There are some papers about this subject in literature. For example, in [6], the author studied dual Fibonacci octonions. In [2], the authors investigated the dual k-Pell octonions. Our aim is to generalize all second order recursive relations over the dual octonions. To achieve this goal, we shall first give some fundamental properties of the Horadam sequence. The Horadam sequence is a particular type of linear recurrence sequences. This sequence generalizes all of the well-known sequences such as Fibonacci, Lucas, Pell, Pell-Lucas sequences, etc. Some studies containing the Horadam sequence can be seen in the references [8–10].

Let us start with the definition of Horadam sequence. In [9], Horadam gave the following sequence definition.

$$\{w_n(a,b;p,q)\}; \quad n \geq 2 \tag{2.1}$$

where

$$w_n = pw_{n-1} + qw_{n-2}; \quad w_0 = a, w_1 = b \tag{2.2}$$

If one writes $a = 0, b = p = q = 1$ in the equation (2.2), then the classical Fibonacci sequence is obtained. Also, the other types of sequences can be obtained such as Lucas, Pell, Pell-Lucas. Now, we present some formulas which will be used very frequently. It is known one of the fundamental identity related to the recurrence relations is the Binet formula, and it can be calculated by using the characteristic equation. For the Horadam sequence, the roots of the characteristic equation are

$$\alpha = \frac{p + \sqrt{p^2 + 4q}}{2}, \quad \beta = \frac{p - \sqrt{p^2 + 4q}}{2}. \tag{2.3}$$

Using these roots and the recurrence relation the Binet formula can be given as follows;

$$w_n = \frac{A\alpha^n - B\beta^n}{\alpha - \beta}, \qquad A = b - a\beta \text{ and } B = b - a\alpha. \tag{2.4}$$

We will examine Binet formula, Cassini identity, d'Ocagne's identity and other equalities for the dual Horadam octonions. It is worth noting that there are some studies related to the second order recurrence relations over $\mathbb{R}, \mathbb{C}, \mathbb{H}$ and $\mathbb{O}$ in literature. One of the earliest studies on this topic belongs to Horadam where it was given a generalization of the second order recurrence relations on $\mathbb{R}$ and $\mathbb{C}$. In [12], Swamy defined the generalized Fibonacci quaternions and then, in [7], we defined quaternions with coefficients from Horadam sequence. Furthermore, in [11], we described and studied the Horadam octonions. And for these octonions, we obtained the Binet formula and some important identities.

3 Dual Horadam octonions

In this section, we define the dual Horadam octonions and give the identities and equalities mentioned in Section 2. Furthermore, we show that the identities involving dual Horadam octonions can be reduced to the identities which are obtained by using different initial values in [6, 13].

Now, let us define a new type of octonions which we call as dual Horadam octonions. Note that these octonions are dual octonions which have coefficients from Horadam sequence. In the rest of the study we will use the notation $\widehat{\mathbb{O}G_n}$ to demonstrate the n-th dual Horadam octonion. Let us define the n-th dual Horadam octonion as follows.

$$\widehat{\mathbb{O}G_n} = (\mathbb{O}G_n, \mathbb{O}G_{n+1}) = \mathbb{O}G_n + \mathbb{O}G_{n+1}\varepsilon \tag{3.1}$$

where

$$\mathbb{O}G_n = w_n e_0 + w_{n+1} e_1 + \cdots + w_{n+7} e_7. \tag{3.2}$$

Firstly, we give some fundamental properties of these dual octonions. For one of the main features of the dual Horadam octonions we give Binet formula in the following theorem.

Theorem 3.1. For $n \geq 1$, we get

$$\widehat{\mathbb{O}G_n} = \frac{A\underline{\alpha}\alpha^n(1+\alpha\varepsilon) - B\underline{\beta}\beta^n(1+\beta\varepsilon)}{\alpha - \beta}, \tag{3.3}$$

where $\underline{\alpha} = 1e_0 + \alpha e_1 + \alpha^2 e_2 + \cdots + \alpha^7 e_7$ and $\underline{\beta} = 1e_0 + \beta e_1 + \beta^2 e_2 + \cdots + \beta^7 e_7$.

Proof. In [11], the Binet formula for Horadam octonions is given as follows.

$$\mathbb{O}G_n = \frac{A\underline{\alpha}\alpha^n - B\underline{\beta}\beta^n}{\alpha - \beta} \tag{3.4}$$

where

$$\underline{\alpha} = 1e_0 + \alpha e_1 + \alpha^2 e_2 + \cdots + \alpha^7 e_7 \text{ and } \underline{\beta} = 1e_0 + \beta e_1 + \beta^2 e_2 + \cdots + \beta^7 e_7. \tag{3.5}$$

For the dual Horadam octonions Binet formula can be obtained in a similar way. Using the definition of dual Horadam octonions and its characteristic equation

$$\widehat{\mathbb{O}G_n} = \frac{A\underline{\alpha}\alpha^n(1+\alpha\varepsilon) - B\underline{\beta}\beta^n(1+\beta\varepsilon)}{\alpha - \beta}$$

is obtained which is desired. Where

$$A = b - a\beta, \quad B = b - a\alpha$$

and

$$\alpha = \frac{p + \sqrt{p^2 + 4q}}{2}, \quad \beta = \frac{p - \sqrt{p^2 + 4q}}{2}.$$

Q.E.D.

One can realize that the equation (3.3) generalizes the Binet formulas which are obtained in [6,13]. That is, for α, β and $A = B = 1$, the formula (3.3) is reduced to following equation.

$$\frac{\underline{\alpha}\alpha^n(1+\alpha\varepsilon) - \underline{\beta}\beta^n(1+\beta\varepsilon)}{\alpha-\beta}.$$

In the next theorem, we give the generating function for dual Horadam octonions.

Theorem 3.2. The generating function for the dual Horadam octonions is

$$\frac{\widehat{\mathbb{O}G_0} + (\widehat{\mathbb{O}G_1} - p\widehat{\mathbb{O}G_0})t}{1-pt-qt^2}. \tag{3.6}$$

Proof. To prove this claim, firstly we write $g(t)$ as

$$g(t) = \widehat{\mathbb{O}G_0}t^0 + \widehat{\mathbb{O}G_1}t + \cdots + \widehat{\mathbb{O}G_n}t^n + \ldots. \tag{3.7}$$

Secondly, we need to calculate $ptg(t)$ and $qt^2g(t)$ in the following equations;

$$ptg(t) = \sum_{n=0}^{\infty} p\widehat{\mathbb{O}G_n}t^{n+1} \text{ and } qt^2g(t) = \sum_{n=0}^{\infty} q\widehat{\mathbb{O}G_n}t^{n+2}. \tag{3.8}$$

Hence, after straightforward calculations, we obtain

$$g(t) = \frac{\mathbb{O}G_0 + \mathbb{O}G_1\varepsilon + (\mathbb{O}G_1 + \mathbb{O}G_2\varepsilon - p(\mathbb{O}G_0 + \mathbb{O}G_1\varepsilon))t}{1-pt-qt^2}, \tag{3.9}$$

which is the generating function for the dual Horadam octonions. Q.E.D.

We want to note that the formula (3.6) generalizes the next formula which is given in [6].

$$\frac{\widehat{\mathbb{O}_0} + (\widehat{\mathbb{O}_1} - \widehat{\mathbb{O}_0})t}{1-t-t^2}. \tag{3.10}$$

For the dual Fibonacci and dual Lucas octonions, Cassini identities are studied in [6,13]. In the following theorem, we give the Cassini formula for dual Horadam octonions.

Theorem 3.3. For dual Horadam octonions, we have

$$\widehat{\mathbb{O}G_{n-1}}\widehat{\mathbb{O}G_{n+1}} - \widehat{\mathbb{O}G_n^2} = \frac{AB(-q)^{n-1}\underline{\alpha}\underline{\beta}[p+(p^2+q)\varepsilon]}{\sqrt{p^2+4q}}. \tag{3.11}$$

Proof. For the proof equation (3.11) if we use the Binet formula, then we have

$$\Big(\frac{A\underline{\alpha}\alpha^{n-1}(1+\alpha\varepsilon) - B\underline{\beta}\beta^{n-1}(1+\beta\varepsilon)}{\alpha-\beta}\Big)\Big(\frac{A\underline{\alpha}\alpha^{n+1}(1+\alpha\varepsilon) - B\underline{\beta}\beta^{n+1}(1+\beta\varepsilon)}{\alpha-\beta}\Big)$$
$$-\Big(\frac{A\underline{\alpha}\alpha^{n}(1+\alpha\varepsilon) - B\underline{\beta}\beta^{n}(1+\beta\varepsilon)}{\alpha-\beta}\Big)^2.$$

After making necessary calculations

$$\frac{AB(-q)^{n-1}\underline{\alpha\beta}[(\alpha^2-\beta^2)+(\alpha^3-\beta^3)\varepsilon]}{(\alpha-\beta)^2}$$

is obtained which can be rearranged as,

$$\frac{(b^2-abp-qa^2)(-q)^{n-1}\underline{\alpha\beta}[p+(p^2+q)\varepsilon]}{\sqrt{p^2+4q}}.$$

Q.E.D.

It should be noted that, in equation (3.11), if we take the necessary initial values then we have a formula which is equivalent to the Cassini formula is given in the reference [6]. Also, according to $\underline{\alpha}$ and $\underline{\beta}$ values for real quaternions, we have Cassini identity stated in [7].

Now, we give d'Ocagne's identity for dual Horadam octonions.

Theorem 3.4. For the dual Horadam octonions, we have

$$\widehat{\mathbb{O}G_{m+1}}\widehat{\mathbb{O}G_n}-\widehat{\mathbb{O}G_m}\widehat{\mathbb{O}G_{n+1}}=\frac{AB\underline{\alpha\beta}(\alpha^n\beta^m-\alpha^m\beta^n)[1+p\varepsilon]}{\sqrt{p^2-4q}}. \tag{3.12}$$

Proof. We use the Binet formula in (3.12). Then the right hand side is follows;

$$=\Big(\frac{A\underline{\alpha}\alpha^{m+1}(1+\alpha\varepsilon)-B\underline{\beta}\beta^{m+1}(1+\beta\varepsilon)}{\alpha-\beta}\Big)\Big(\frac{A\underline{\alpha}\alpha^{n}(1+\alpha\varepsilon)-B\underline{\beta}\beta^{n}(1+\beta\varepsilon)}{\alpha-\beta}\Big)$$
$$-\Big(\frac{A\underline{\alpha}\alpha^{m}(1+\alpha\varepsilon)-B\underline{\beta}\beta^{m}(1+\beta\varepsilon)}{\alpha-\beta}\Big)\Big(\frac{A\underline{\alpha}\alpha^{n+1}(1+\alpha\varepsilon)-B\underline{\beta}\beta^{n+1}(1+\beta\varepsilon)}{\alpha-\beta}\Big).$$

After needed calculations,

$$\widehat{\mathbb{O}G_{m+1}}\widehat{\mathbb{O}G_n}-\widehat{\mathbb{O}G_m}\widehat{\mathbb{O}G_{n+1}}=\frac{AB\underline{\alpha\beta}(\alpha^n\beta^m-\alpha^m\beta^n)[1+p\varepsilon]}{\sqrt{p^2-4q}}$$

is obtained where $AB=b^2-abp-qa^2$. Q.E.D.

Note that if we focus on non-dual part of the equation (3.12) we get d'Ocagne's identity for real Horadam octonions.

Remark 3.5. For real Horadam octonions, d'Ocagne's identity is as follows;

$$\mathbb{O}G_{m+1}\mathbb{O}G_n-\mathbb{O}G_m\mathbb{O}G_{n+1}=\frac{AB\underline{\alpha\beta}(\alpha^n\beta^m-\alpha^m\beta^n)}{\alpha-\beta}.$$

In the following theorem, we give Vajda identity for dual Horadam octonions.

Theorem 3.6. For the dual Horadam octonions, we have

$$\widehat{\mathbb{O}G_{n+m}}\widehat{\mathbb{O}G_{n+k}}-\widehat{\mathbb{O}G_{n}}\widehat{\mathbb{O}G_{n+m+k}}=$$
$$\frac{AB\underline{\alpha}\underline{\beta}(-q)^{n}\beta^{k}(\beta^{k}-\alpha^{m})+(\beta^{m+1}+\alpha\beta^{m}-\alpha^{m+1}-\alpha^{m}\beta)\varepsilon}{(\alpha-\beta)^{2}}+$$
$$\frac{AB\underline{\beta}\underline{\alpha}(-q)^{n}\alpha^{k}(\alpha^{k}-\beta^{m})+(\alpha^{m+1}+\alpha^{m}\beta-\beta^{m+1}-\beta^{m}\alpha)\varepsilon}{(\alpha-\beta)^{2}}. \tag{3.13}$$

Proof. We use the equation (3.3) and rewrite the equation (3.13) explicitly as follows

$$=\Big(\frac{A\underline{\alpha}\alpha^{n+m}(1+\alpha\varepsilon)-B\underline{\beta}\beta^{n+m}(1+\beta\varepsilon)}{\alpha-\beta}\Big)\Big(\frac{A\underline{\alpha}\alpha^{n+k}(1+\alpha\varepsilon)-B\underline{\beta}\beta^{n+k}(1+\beta\varepsilon)}{\alpha-\beta}\Big)$$
$$-\Big(\frac{A\underline{\alpha}\alpha^{n}(1+\alpha\varepsilon)-B\underline{\beta}\beta^{n}(1+\beta\varepsilon)}{\alpha-\beta}\Big)\Big(\frac{A\underline{\alpha}\alpha^{n+m+k}(1+\alpha\varepsilon)-B\underline{\beta}\beta^{n+m+k}(1+\beta\varepsilon)}{\alpha-\beta}\Big).$$

After needed calculations Vajda identity is

$$\widehat{\mathbb{O}G_{n+m}}\widehat{\mathbb{O}G_{n+k}}-\widehat{\mathbb{O}G_{n}}\widehat{\mathbb{O}G_{n+m+k}}=$$
$$\frac{AB\underline{\alpha}\underline{\beta}\alpha^{n}\beta^{n+k}(\beta^{k}-\alpha^{m})+(\beta^{m+1}+\alpha\beta^{m}-\alpha^{m+1}-\alpha^{m}\beta)\varepsilon}{(\alpha-\beta)^{2}}+$$
$$\frac{AB\underline{\beta}\underline{\alpha}\alpha^{n+k}\beta^{n}(\alpha^{k}-\beta^{m})+(\alpha^{m+1}+\alpha^{m}\beta-\beta^{m+1}-\beta^{m}\alpha)\varepsilon}{(\alpha-\beta)^{2}}$$

where $AB=b^{2}-abp-qa^{2}$. Q.E.D.

For the real Horadam octonions we give following remark.

Remark 3.7. For the real Horadam octonions the Vajda identity is

$$\mathbb{O}G_{n+m}\mathbb{O}G_{n+k}-\mathbb{O}G_{n}\mathbb{O}G_{n+m+k}=\frac{AB\underline{\alpha}\underline{\beta}(-q)^{n}(\beta^{2k}+\beta^{n+m}-\alpha^{2k}-\alpha^{m+k})}{(\alpha-\beta)^{2}}. \tag{3.14}$$

Theorem 3.8. For dual Horadam octonions, we have the following relations:

$$i)\ \widehat{\mathbb{O}G_{n}}\widehat{\mathbb{O}G_{n+3}}-\widehat{\mathbb{O}G_{n+1}}\widehat{\mathbb{O}G_{n+2}}=-AB\underline{\alpha}\underline{\beta}(-q)^{n}[p+(p^{2}+2q)\varepsilon]. \tag{3.15}$$

$$ii)\ \widehat{\mathbb{O}G^{2}_{n}}\pm\widehat{\mathbb{O}G^{2}_{n+1}}=$$
$$A^{2}\underline{\alpha}^{2}\alpha^{2n}[(1\pm\alpha^{2})+2(\alpha\pm\alpha^{3})\varepsilon]+B^{2}\underline{\beta}^{2}\beta^{2n}[(1\pm\beta^{2})+2(\beta\pm\beta^{3})\varepsilon]. \tag{3.16}$$

Proof. Let us prove the equation (3.16) with plus sign which is

$$\widehat{\mathbb{O}G^{2}_{n}}+\widehat{\mathbb{O}G^{2}_{n+1}}=A^{2}\underline{\alpha}^{2}\alpha^{2n}[(1+\alpha^{2})+2(\alpha+\alpha^{3})\varepsilon]+B^{2}\underline{\beta}^{2}\beta^{2n}[(1+\beta^{2})+2(\beta+\beta^{3})\varepsilon].$$

In order to prove we use the equation (3.3) in the above equation as

$$\Big(\frac{A\underline{\alpha}\alpha^{n}(1+\alpha\varepsilon)-B\underline{\beta}\beta^{n}(1+\beta\varepsilon)}{\alpha-\beta}\Big)^{2}+\Big(\frac{A\underline{\alpha}\alpha^{n+1}(1+\alpha\varepsilon)-B\underline{\beta}\beta^{n+1}(1+\beta\varepsilon)}{\alpha-\beta}\Big)^{2}.$$

After using properties of α and β we get

$$A^2\underline{\alpha}^2\alpha^{2n}[(1+\alpha^2)+2(\alpha+\alpha^3)\varepsilon]+B^2\underline{\beta}^2\beta^{2n}[(1+\beta^2)+2(\beta+\beta^3)\varepsilon]$$

which is desired. Q.E.D.

In analogy with the Theorem 3 we state following remark.

Remark 3.9. For real Horadam octonions we have the following identities

$$i)\ \mathbb{O}G_n\mathbb{O}G_{n+3}-\mathbb{O}G_{n+1}\mathbb{O}G_{n+2}=-AB\underline{\alpha}\underline{\beta}(-q)^np. \tag{3.17}$$

$$ii)\ \mathbb{O}G_n^2\pm\mathbb{O}G_{n+1}^2=A^2\underline{\alpha}^2\alpha^{2n}(1\pm\alpha^2)+B^2\underline{\beta}^2\beta^{2n}(1\pm\beta^2). \tag{3.18}$$

In the following theorem, we give a formula which gives the sum of dual Horadam octonions

Theorem 3.10. The summation formula for the dual Horadam octonions is

$$\sum_{i=0}^{n}\widehat{\mathbb{O}G_i}=\sum_{i=0}^{n}\mathbb{O}G_i+\sum_{i=1}^{n+1}\mathbb{O}G_i\varepsilon=(d_1,d_2), \tag{3.19}$$

where the real and dual parts of equation (3.19) are as follows:

$$d_1=\sum_{i=0}^{n}\mathbb{O}G_i=\frac{1}{\alpha-\beta}\left(\frac{B\underline{\beta}\beta^{n+1}}{1-\beta}-\frac{A\underline{\alpha}\alpha^{n+1}}{1-\alpha}\right)+K \tag{3.20}$$

and

$$d_2=\sum_{i=1}^{n+1}\mathbb{O}G_i=\frac{1}{\alpha-\beta}\left(\frac{B\underline{\beta}\beta^{n+2}}{1-\beta}-\frac{A\underline{\alpha}\alpha^{n+2}}{1-\alpha}\right)-\mathbb{O}G_0+K \tag{3.21}$$

respectively. The value K is

$$K=\frac{A\underline{\alpha}(1-\beta)-B\underline{\beta}(1-\alpha)}{(\alpha-\beta)(1-\alpha)(1-\beta)}. \tag{3.22}$$

Proof. For the summation formula of $\widehat{\mathbb{O}G_n}$ we calculate the values d_1 and d_2 separately. To find d_1 we can use Binet formula and definition of geometric series as follows:

$$d_1=\sum_{i=0}^{n}\mathbb{O}G_i=\frac{A\underline{\alpha}\alpha^n-B\underline{\beta}\beta^n}{\alpha-\beta}=\frac{A\underline{\alpha}}{\alpha-\beta}\sum_{i=0}^{n}\alpha^n-\frac{B\underline{\beta}}{\alpha-\beta}\sum_{i=0}^{n}\beta^n.$$

$$=\frac{A\underline{\alpha}}{\alpha-\beta}\sum_{i=0}^{n}\frac{1-\alpha^{n+1}}{1-\alpha}-\frac{B\underline{\beta}}{\alpha-\beta}\sum_{i=0}^{n}\frac{1-\beta^{n+1}}{1-\beta}.$$

Upon making the necessary calculations, we get the explicit form of d_1 as

$$\sum_{i=0}^{n}\mathbb{O}G_i=\frac{1}{\alpha-\beta}\left(\frac{B\underline{\beta}\beta^{n+1}}{1-\beta}-\frac{A\underline{\alpha}\alpha^{n+1}}{1-\alpha}\right)+\frac{A\underline{\alpha}(1-\beta)-B\underline{\beta}(1-\alpha)}{(\alpha-\beta)(1-\alpha)(1-\beta)}.$$

On the other hand, the value of d_2 can easily be calculated similar to d_1. Q.E.D.

We note that the formula in the equation (3.19) generalizes the formula is given by Halici in [6], which is for dual Fibonacci octonions;

$$\sum_{i=1}^{n} \widehat{\mathbb{O}_i} = \widehat{\mathbb{O}_2} F_{n+1} + \widehat{\mathbb{O}_1} F_{n-1} - \widehat{\mathbb{O}_2}. \tag{3.23}$$

Theorem 3.11. The norm of *n-th* dual Horadam octonion is

$$Nr(\widehat{\mathbb{O}G_n}) = (\mathbb{O}G_n \overline{\mathbb{O}G_n},\ \mathbb{O}G_n \overline{\mathbb{O}G_{n+1}} + \mathbb{O}G_{n+1} \overline{\mathbb{O}G_n}) = (e_1, e_2) \tag{3.24}$$

where e_1, e_2 and L are follows:

$$e_1 = \frac{A^2\alpha^{2n}(1+\alpha^2+\alpha^4+\cdots+\alpha^{14}) + B^2\beta^{2n}(1+\beta^2+\beta^4+\cdots+\beta^{14})}{(\alpha-\beta)^2} - L, \tag{3.25}$$

$$e_2 = 2(\sum_{i=0}^{7} w_{n+i} w_{n+1+i}) \tag{3.26}$$

and

$$L = \frac{2AB(-q)^n(a+(-q)+\cdots+(-q)^7)}{p^2+4q} \tag{3.27}$$

where $AB = b^2 - abp - qa^2$.

Proof. It follows from the definition of norm that,

$$Nr(\widehat{\mathbb{O}G_n}) = (\mathbb{O}G_n \overline{\mathbb{O}G_n},\ \mathbb{O}G_n \overline{\mathbb{O}G_{n+1}} + \mathbb{O}G_{n+1} \overline{\mathbb{O}G_n}).$$

To get the desired results we first calculate e_1 as follows:

$$e_1 = \frac{A^2\alpha^{2n}(1+\alpha^2+\alpha^4+\cdots+\alpha^{14})}{(\alpha-\beta)^2} + \frac{B^2\beta^{2n}(1+\beta^2+\beta^4+\cdots+\beta^{14})}{(\alpha-\beta)^2} - \frac{2AB(-q)^n(a+(-q)+\cdots+(-q)^7)}{(\alpha-\beta)^2}. \tag{3.28}$$

The dual part of the equation (3.24) can easily be calculated because of the property $\varepsilon^2 = 0$. After direct calculations, the following relation is obtained.

$$e_2 = \mathbb{O}G_n \overline{\mathbb{O}G_{n+1}} + \mathbb{O}G_{n+1} \overline{\mathbb{O}G_n} = 2\Big(\sum_{i=0}^{7} w_{n+i} w_{n+1+i}\Big). \tag{3.29}$$

Q.E.D.

Norm values of dual Fibonacci and Lucas octonions are studied in [6,13]. Utilizing the equation (3.24) together with the appropriate initial values one can get the following equation which is identical to the norm of dual Fibonacci octonions is given in [6]

$$Nr(\widehat{\mathbb{O}_n}) = 21(F_{2n+7} + 2F_{2n+8}\varepsilon). \tag{3.30}$$

In order to see that 3.24 contains real Horadam quaternions (see [7]) one can calculate the $\underline{\alpha}$, $\underline{\beta}$ and use real part of the equation.

4 Conclusion

This paper generalizes similar papers in this area up to now on dual quaternions and octonions. Thanks to this generalization the important properties concerning all the dual octonions such as the Binet formula, the generating function, the Cassini, d'Ocagne's and Vadja identities can be easily obtained.

References

[1] J. Baez, *The octonions*, Bulletin of the American Mathematical Society **(39)2** (2002) 145–205.

[2] P. Catarino and P. Vasco, *On dual k-pell quaternions and octonions*, Mediterranean Journal of Mathematics, **(14)2** (2017) 75.

[3] J. H. Conway and D. A. Smith, *On quaternions and octonions*, AK Peters/CRC Press (2003).

[4] L. E. Dickson, *On quaternions and their generalization and the history of the eight square theorem*, Annals of Mathematics (1919) 155–171.

[5] C. Flaut and V. Shpakivskyi, *Some identities in algebras obtained by the cayley-dickson process*, Advances in Applied Clifford Algebras **(23)1** (2013) 63–76.

[6] S. Halıcı, *On dual fibonacci octonions*, Advances in Applied Clifford Algebras **(25)4** (2015) 905–914.

[7] S. Halıcı and A. Karataş, *On a generalization for fibonacci quaternions*, Chaos, Solitons and Fractals **98** (2017) 178–182.

[8] P. Haukkanen, *A note on horadam's sequence*, Fibonacci Quart. **(40)4** (2002) 358–361.

[9] A. F. Horadam, *Basic properties of a certain generalized sequence of numbers.*, Fibonacci Quart. **(3)3** (1965) 161–176.

[10] A. F. Horadam, *Special properties of the sequence wn (a, b; p, q)*, Fibonacci Quart. **(5)5** (1967) 424–434.

[11] A. Karataş and S. Halici, *Horadam octonions*, Analele Universitatii" Ovidius" Constanta-Seria Matematica **(25)3** (2017) 97–106.

[12] M. N. S. Swamy, *On generalized fibonacci quaternions*, Fibonacci Quart. **(11)5** (1973) 547–549.

[13] Z. Űnal, Ű. Tokeser and G. Bilgici, *Some properties of dual fibonacci and dual lucas octonions*, Advances in Applied Clifford Algebras **27.2** (2016) 1907–1916.

DOI 10.2478/9788395793882-007

On the jerk in motion along a regular surface curve

Murat Tosun* and Merve Hızarcıoğlu

Department of Mathematics, Faculty of Arts and Sciences, Sakarya University, Sakarya, Turkey
*Corresponding Author
E-mail: tosun@sakarya.edu.tr*, maturk54@gmail.com

Abstract

The jerk is the third time derivative of the position vector, and hence the time derivative of acceleration vector. In this paper, we take into consideration a particle moving on a regular surface curve, which is equipped with the Darboux frame, in three dimensional Euclidean space and resolve its jerk vector along tangential direction and two special radial directions. Moreover, the motion of a particle along a right handed circular helix is given as an illustrative example.

2010 Mathematics Subject Classification. **70B05** 14H50
Keywords. kinematics of a particle, plane and space curves, jerk, Darboux frame, regular surface curve.

1 Introduction

The force acting on a particle is concerned with its acceleration through the equality $\mathbf{F} = m\mathbf{a}$ in Newtonian physics. The jerk $\mathbf{J}$ is the time derivative of acceleration. Thus, if the mass is constant the equality $\mathbf{J} = \frac{1}{m}\frac{d\mathbf{F}}{dt}$ holds. If the time derivative of force is nonzero, the jerk vector is a nonzero vector.

For a large acceleration of short duration, the jerk vector has precious applications in the design of intermittent-motion mechanisms, e.g. cams and genevas [1,2]. When a gymnast does gymnastic exercises or a machinist drives a high-speed train or a stock-car racer races on track, the acceleration changes suddenly. In situations like this, to estimate the lower threshold of just noticeable jerk and upper values of the jerk that can be tolerated by humans without undue discomfort has an important place [3]. These calculations were considered by Melchior in [4].

The decomposition of the jerk vector along a curve in 3-dimensional Euclidean space E^3 is well known in the literature due to the study [5] of Resal. In this decomposition, the jerk vector lies along the tangential, normal and binormal components. Recently, a new resolution of jerk vector along the tangential direction, radial direction in osculating plane and the other radial direction in rectifying plane has been presented by Özen et al. [6]. Also, more recently, the jerk of a particle moving on an analytic space curve equipped with the modified orthogonal frame has been investigated by Özen et al. [7]. In the present paper, we inspire from the studies [6, 7] and discuss the same topic according to Darboux frame for a particle moving on a regular surface curve.

In this paper, firstly, we have given a short knowledge about the Serret-Frenet frame and Darboux frame. Afterwards, for a particle moving along a regular surface curve, which is equipped with the Darboux frame, we have obtained the resolution of the jerk vector along tangential and radial directions. Moreover, an example for circular helices has been given.

Tbilisi Mathematical Journal Special Issue (IECMSA-2019), pp. 69–77.
Tbilisi Centre for Mathematical Sciences.

Received by the editors: 03 October 2019.
Accepted for publication: 02 January 2020.

2 Preliminaries

Let E^3 be equipped with the standard scalar product

$$\mathbf{U}\cdot\mathbf{V}=u_1v_1+u_2v_2+u_3v_3$$

where $\mathbf{U}=(u_1\,,\,u_2\,,\,u_3)\,,\,\mathbf{V}=(v_1\,,\,v_2\,,\,v_3)$ are any vectors in E^3. The norm of a vector $\mathbf{U}\in E^3$ is defined as $\|\mathbf{U}\|=\sqrt{\mathbf{U}\cdot\mathbf{U}}$. A curve $\alpha=\alpha(s):I\subset R\to E^3$ is called a unit speed curve if $\|\alpha'(s)\|=1$ for all $s\in I$. In that case, s is called arc-length parameter of $\alpha(s)$.

The Serret-Frenet frame of the curve $\alpha(s)$ is denoted by $\{\mathbf{T}(s)\,,\,\mathbf{N}(s)\,,\,\mathbf{B}(s)\}$ where the unit vectors $\mathbf{T}(s)\,,\ \mathbf{N}(s)$ and $\mathbf{B}(s)$ are the unit tangent, unit principal normal and unit binormal vectors respectively. Also, the Serret-Frenet formulas are given as follows:

$$\mathbf{T}'=\kappa\mathbf{N},\qquad \mathbf{N}'=-\kappa\mathbf{T}+\tau\mathbf{B},\qquad \mathbf{B}'=-\tau\mathbf{N} \tag{2.1}$$

where $\kappa(s)=\|\mathbf{T}'(s)\|$ is the curvature function and $\tau(s)=-(\mathbf{B}'(s)\cdot\mathbf{N}(s))$ is the torsion function [8].

Let M be a regular surface and $\alpha:I\subset R\to M$ be a unit speed curve on this surface. In this case, there exists the Darboux frame which is showed with $\{\mathbf{T},\,\mathbf{Y},\,\mathbf{U}\}$ along the curve α. In this frame, $\mathbf{T}$ is the unit tangent vector of the curve α, and $\mathbf{U}$ is the unit normal vector of the surface M restricted to the curve α. On the other hand, $\mathbf{Y}$ is the unit vector which is obtained by vector product these two vectors, that is, $\mathbf{Y}=\mathbf{U}\times\mathbf{T}$. Because the unit tangent vector $\mathbf{T}$ is mutual in both Serret-Frenet frame and Darboux frame, the vectors $\mathbf{N}$, $\mathbf{B}$, $\mathbf{Y}$, $\mathbf{U}$ lie in the same plane. Thus, there is a relation between these frames as in the following:

$$\mathbf{T}=\mathbf{T},\qquad \mathbf{Y}=\cos\varphi\,\mathbf{N}-\sin\varphi\,\mathbf{B},\qquad \mathbf{U}=\sin\varphi\,\mathbf{N}+\cos\varphi\,\mathbf{B} \tag{2.2}$$

where φ is the angle between the vectors $\mathbf{U}$ and $\mathbf{B}$ (or $\mathbf{Y}$ and $\mathbf{N}$). Furthermore, the derivative formulas of the Darboux frame are given by

$$\mathbf{T}'=k_g\mathbf{Y}+k_n\mathbf{U},\qquad \mathbf{Y}'=-k_g\mathbf{T}+\tau_g\mathbf{U},\qquad \mathbf{U}'=-k_n\mathbf{T}-\tau_g\mathbf{Y} \tag{2.3}$$

where k_n is the normal curvature, k_g is the geodesic curvature and τ_g is the geodesic torsion of α. Also, k_g, k_n and τ_g satisfy the equations

$$\begin{aligned} k_g(s) &= \sqrt{{k_g}^2(s)+{k_n}^2(s)}\cos\varphi(s)\,,\\ k_n(s) &= \sqrt{{k_g}^2(s)+{k_n}^2(s)}\sin\varphi(s)\,,\\ \tau_g(s) &= \tau(s)-\frac{d\varphi(s)}{ds} \end{aligned} \tag{2.4}$$

where $\varphi(s)=\arctan\left(k_n(s)/k_g(s)\right)$ and $\sqrt{{k_g}^2(s)+{k_n}^2(s)}=\kappa(s)$ [9–12].

Let a particle P of mass $m\,(>0)$ move on a regular surface M in E^3. Choose any fixed origin O in the space and symbolize the position vector of P at time t by $\mathbf{x}$. Let C, with the arc length parameter s, be the oriented curve traced out by P where the arc length of C corresponds t. In that case, the unit tangent vector for the curve C can be written as

$$\mathbf{T}=\frac{d\mathbf{x}}{ds}. \tag{2.5}$$

By means of (2.1) and (2.5), the velocity vector $\mathbf{v} = d\mathbf{x}/dt$, the acceleration vector $\mathbf{a} = d\mathbf{v}/dt$ and the jerk vector $\mathbf{J} = d\mathbf{a}/dt$ of P at time t are given by (see [13] for more details):

$$\begin{aligned}
\mathbf{v} &= \left(\frac{ds}{dt}\right)\mathbf{T}, \\
\mathbf{a} &= \left(\frac{d^2s}{dt^2}\right)\mathbf{T} + \kappa\left(\frac{ds}{dt}\right)^2\mathbf{N}, \\
\mathbf{J} &= \left[\left(\frac{d^3s}{dt^3}\right) - \left(\frac{ds}{dt}\right)^3\kappa^2\right]\mathbf{T} + \left[3\left(\frac{ds}{dt}\right)\left(\frac{d^2s}{dt^2}\right)\kappa + \left(\frac{ds}{dt}\right)^3\frac{d\kappa}{ds}\right]\mathbf{N} + \left[\kappa\tau\left(\frac{ds}{dt}\right)^3\right]\mathbf{B}.
\end{aligned}$$

On the other hand, the acceleration vector $\mathbf{a}$ of P at time t is given as

$$\mathbf{a} = \left(\frac{d^2s}{dt^2}\right)\mathbf{T} + \sqrt{{k_g}^2 + {k_n}^2}\left(\frac{ds}{dt}\right)^2\cos\varphi\,\mathbf{Y} + \sqrt{{k_g}^2 + {k_n}^2}\left(\frac{ds}{dt}\right)^2\sin\varphi\,\mathbf{U} \tag{2.6}$$

with respect to Darboux basis [14].

3 Jerk of a particle moving along a regular surface curve

In this section, the jerk vector of a particle moving along a regular surface curve is given according to Darboux basis. Moreover, the resolution of the jerk vector along tangential and two special radial directions has been found. To do so, we continue to take into consideration the aforesaid particle P.

By using (2.3), (2.4) and (2.6), we can immediately get the jerk vector $\mathbf{J}$ of P at time t with respect to Darboux basis as follows:

$$\mathbf{J} = C_{\mathbf{T}}\,\mathbf{T} + C_{\mathbf{Y}}\,\mathbf{Y} + C_{\mathbf{U}}\mathbf{U} \tag{3.1}$$

where

$$\begin{aligned}
C_{\mathbf{T}} &= \left(\frac{d^3s}{dt^3}\right) - \left(\frac{ds}{dt}\right)^3\left({k_g}^2 + {k_n}^2\right), \\
C_{\mathbf{Y}} &= \cos\varphi\left[3\left(\frac{ds}{dt}\right)\left(\frac{d^2s}{dt^2}\right)\sqrt{{k_g}^2 + {k_n}^2} + \frac{d\left(\sqrt{{k_g}^2 + {k_n}^2}\right)}{ds}\left(\frac{ds}{dt}\right)^3\right] \\
&\quad - \sin\varphi\left[\sqrt{{k_g}^2 + {k_n}^2}\left(\tau_g + \frac{d\varphi}{ds}\right)\left(\frac{ds}{dt}\right)^3\right], \\
C_{\mathbf{U}} &= \sin\varphi\left[3\left(\frac{ds}{dt}\right)\left(\frac{d^2s}{dt^2}\right)\sqrt{{k_g}^2 + {k_n}^2} + \frac{d\left(\sqrt{{k_g}^2 + {k_n}^2}\right)}{ds}\left(\frac{ds}{dt}\right)^3\right] \\
&\quad + \cos\varphi\left[\sqrt{{k_g}^2 + {k_n}^2}\left(\tau_g + \frac{d\varphi}{ds}\right)\left(\frac{ds}{dt}\right)^3\right].
\end{aligned}$$

This equation yields

$$\begin{aligned}\mathbf{J} &= \left[\left(\frac{d^3s}{dt^3}\right) - \left(\frac{ds}{dt}\right)^3 \left({k_g}^2 + {k_n}^2\right)\right]\mathbf{T} \\ &+ \left[3\left(\frac{ds}{dt}\right)\left(\frac{d^2s}{dt^2}\right)\sqrt{{k_g}^2 + {k_n}^2} + \frac{d\left(\sqrt{{k_g}^2 + {k_n}^2}\right)}{ds}\left(\frac{ds}{dt}\right)^3\right](\cos\varphi\mathbf{Y} + \sin\varphi\mathbf{U}) \\ &+ \left[\sqrt{{k_g}^2 + {k_n}^2}\left(\tau_g + \frac{d\varphi}{ds}\right)\left(\frac{ds}{dt}\right)^3\right](-\sin\varphi\,\mathbf{Y} + \cos\varphi\,\mathbf{U}).\end{aligned} \tag{3.2}$$

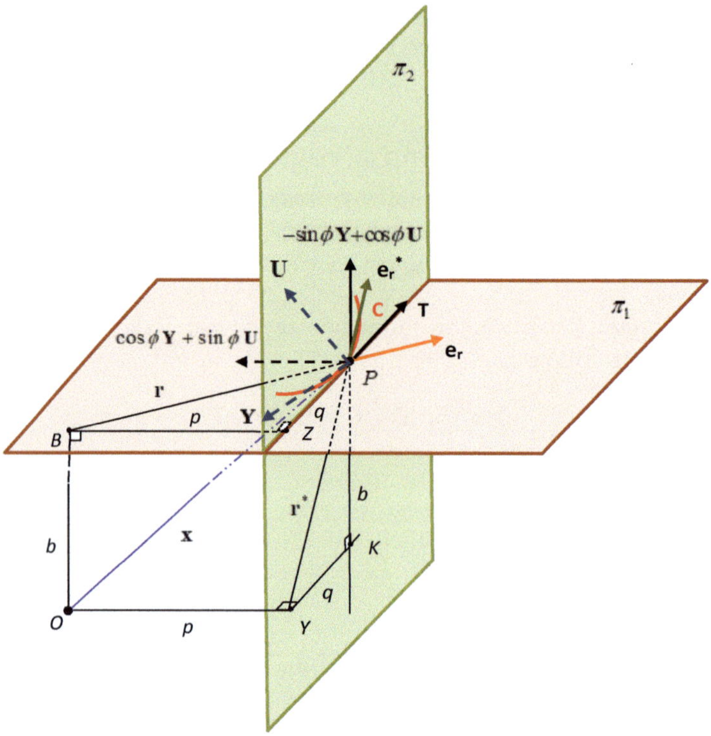

FIGURE 1. A particle P moves along a curve C on a regular surface M in 3-dimensional Euclidean space (To provide visual harmony of the figure, the surface M has been removed from the figure). B is the foot of the perpendicular which is from origin O to the plane π_1 and BZ is perpendicular to the tangent line. The position vector of P relative to B is shown with $\mathbf{r}$, and $\mathbf{e_r}$ is the unit vector in direction of BP. Y is the foot of the perpendicular which is from origin O to the plane π_2 and YK is perpendicular to the line determined by the vector $(-\sin\varphi\,\mathbf{Y} + \cos\varphi\,\mathbf{U})$. The position vector of P relative to Y is shown with $\mathbf{r}^*$, and $\mathbf{e_r}^*$ is the unit vector in direction of YP.

A particle, moving on a curve, may be seen as a point of this curve. So, P has a position vector in terms of the Darboux basis of the curve C. Suppose that the position vector of P on the Darboux basis is resolved as in the following:

$$\mathbf{x} = q\,\mathbf{T} - p\,(\cos\varphi\,\mathbf{Y} + \sin\varphi\,\mathbf{U}) + b\,(-\sin\varphi\,\mathbf{Y} + \cos\varphi\,\mathbf{U}) \tag{3.3}$$

where

$$q = \mathbf{x}\cdot\mathbf{T}, \quad -p = \mathbf{x}\cdot(\cos\varphi\,\mathbf{Y} + \sin\varphi\,\mathbf{U}), \quad b = \mathbf{x}\cdot(-\sin\varphi\,\mathbf{Y} + \cos\varphi\,\mathbf{U}). \tag{3.4}$$

Show with $\mathbf{r}$ and $\mathbf{r}^*$ the vectors

$$\mathbf{r} = q\mathbf{T} - p\,(\cos\varphi\,\mathbf{Y} + \sin\varphi\,\mathbf{U}), \quad \mathbf{r}^* = q\mathbf{T} + b\,(-\sin\varphi\,\mathbf{Y} + \cos\varphi\,\mathbf{U}) \tag{3.5}$$

lying in plane $\pi_1 = Sp\,\{\mathbf{T}, \cos\varphi\mathbf{Y} + \sin\varphi\mathbf{U}\}$ to C at P and in plane $\pi_2 = Sp\,\{\mathbf{T}, -\sin\varphi\mathbf{Y} + \cos\varphi\mathbf{U}\}$ to C at P, respectively. In that case, we get the equalities

$$r^2 = \mathbf{r}\cdot\mathbf{r} = p^2 + q^2, \quad (r^*)^2 = \mathbf{r}^*\cdot\mathbf{r}^* = q^2 + b^2 \tag{3.6}$$

where r and r^* are the lengths of the vectors $\mathbf{r}$ and $\mathbf{r}^*$, respectively (see Figure 1).

By vector product of the position vector $\mathbf{x}$ given in (3.3), and the linear momentum vector $\mathbf{p} = m\left(\frac{ds}{dt}\right)\mathbf{T}$, the angular momentum vector of P about O is obtained as follows:

$$\mathbf{H}^O = m\,b\left(\frac{ds}{dt}\right)(\cos\varphi\,\mathbf{Y} + \sin\varphi\,\mathbf{U}) + mp\left(\frac{ds}{dt}\right)(-\sin\varphi\,\mathbf{Y} + \cos\varphi\,\mathbf{U}). \tag{3.7}$$

By using the equation (3.5), let us express the vector $(\cos\varphi\,\mathbf{Y} + \sin\varphi\,\mathbf{U})$ in terms of $\mathbf{r}$ and $\mathbf{T}$ and express the vector $(-\sin\varphi\,\mathbf{Y} + \cos\varphi\,\mathbf{U})$ in terms of $\mathbf{r}^*$ and $\mathbf{T}$ to resolve the jerk vector $\mathbf{J}$ in (3.2) along the tangential direction, radial direction BP in the plane π_1, and radial direction YP in the plane π_2. But, these are possible if and only if $p \neq 0$ and $b \neq 0$. If we make the physical assumption that the components of the angular momentum along the vectors $(\cos\varphi\,\mathbf{Y} + \sin\varphi\,\mathbf{U})$ and $(-\sin\varphi\,\mathbf{Y} + \cos\varphi\,\mathbf{U})$ never vanish, we can ensure that each of p and b are not equal to zero. Then, we can write the equations

$$\cos\varphi\,\mathbf{Y} + \sin\varphi\,\mathbf{U} = \frac{1}{p}(-\mathbf{r} + q\mathbf{T}), \quad -\sin\varphi\,\mathbf{Y} + \cos\varphi\,\mathbf{U} = \frac{1}{b}(-q\mathbf{T} + \mathbf{r}^*). \tag{3.8}$$

Also, from the aforesaid physical assumption and the equation (3.6), $r \neq 0$ and $r^* \neq 0$. Therefore, the unit vectors $\mathbf{e_r}$ and $\mathbf{e_r}^*$ can be defined as

$$\mathbf{e_r} = \frac{1}{r}\mathbf{r}, \quad \mathbf{e_r}^* = \frac{1}{r^*}\mathbf{r}^*. \tag{3.9}$$

By means of (3.8) and (3.9), we get

$$\cos\varphi\mathbf{Y} + \sin\varphi\mathbf{U} = \frac{(-r\,\mathbf{e_r} + q\mathbf{T})}{p}, \quad -\sin\varphi\mathbf{Y} + \cos\varphi\mathbf{U} = \frac{(-q\mathbf{T} + r^*\mathbf{e_r}^*)}{b}. \tag{3.10}$$

Now, if we substitute the equation (3.10) into the equation (3.2), we obtain the jerk vector $\mathbf{J}$ as in the following:

$$\begin{aligned}
\mathbf{J} &= \left[\begin{array}{c} \left(\frac{d^3s}{dt^3}\right) - \left(\frac{ds}{dt}\right)^3\left({k_g}^2+{k_n}^2\right) + 3\frac{q}{p}\left(\frac{ds}{dt}\right)\left(\frac{d^2s}{dt^2}\right)\sqrt{{k_g}^2+{k_n}^2} \\ +\frac{q}{p}\frac{d\left(\sqrt{{k_g}^2+{k_n}^2}\right)}{ds}\left(\frac{ds}{dt}\right)^3 - \frac{q}{b}\left(\frac{ds}{dt}\right)^3\sqrt{{k_g}^2+{k_n}^2}\left(\tau_g+\frac{d\varphi}{ds}\right) \end{array}\right]\mathbf{T} \\
&\quad + \left[-3\frac{r}{p}\left(\frac{ds}{dt}\right)\left(\frac{d^2s}{dt^2}\right)\sqrt{{k_g}^2+{k_n}^2} - \frac{r}{p}\frac{d\left(\sqrt{{k_g}^2+{k_n}^2}\right)}{ds}\left(\frac{ds}{dt}\right)^3\right]\mathbf{e_r} \qquad (3.11) \\
&\quad + \left[\frac{r^*}{b}\left(\frac{ds}{dt}\right)^3\sqrt{{k_g}^2+{k_n}^2}\left(\tau_g+\frac{d\varphi}{ds}\right)\right]\mathbf{e_r}^* \\
&= T_t\,\mathbf{T} + T_r\,\mathbf{e_r} + T_{r^*}\,\mathbf{e_r}^*.
\end{aligned}$$

Here, T_r, T_{r^*} and T_t are the first radial, second radial and tangential components of the jerk, respectively.

By taking into consideration the expressions and equations which are given above, we can state the following theorem:

Theorem 3.1. Let a particle P of mass $m\,(>0)$ move on a regular surface curve C in 3-dimensional Euclidean space E^3, and suppose that each of the components of angular momentum along the unit vector $(\cos\varphi\,\mathbf{Y}+\sin\varphi\,\mathbf{U})$ and along the unit vector $(-\sin\varphi\,\mathbf{Y}+\cos\varphi\,\mathbf{U})$ never vanish. In that case, the jerk vector $\mathbf{J}$ of the particle P can be expressed as in (3.11).

4 Motion of a particle along a right-handed circular helix

Suppose that a particle P travels along a right-handed circular helix which lies on a cylinder of radius R. The position vector of P is given by

$$\mathbf{x} = R\cos(\omega t)\,\mathbf{i} + R\sin(\omega t)\,\mathbf{j} + v_z t\,\mathbf{k}$$

in Cartesian coordinates where t and ω indicate the time and time independent angular frequency, respectively. Here, $\{\mathbf{i},\mathbf{j},\mathbf{k}\}$ represents a fixed right-handed orthonormal frame, v_z represents a positive constant. Let the axis $\mathbf{k}$ be the axis of the helix and α be the helix angle which satisfies $\tan\alpha = \frac{R\omega}{v_z}$. Then, we get

$$\begin{aligned}
\mathbf{v} &= -R\omega\sin(\omega t)\,\mathbf{i} + R\omega\cos(\omega t)\,\mathbf{j} + v_z\mathbf{k}, \\
\mathbf{a} &= -R\omega^2\cos(\omega t)\,\mathbf{i} - R\omega^2\sin(\omega t)\,\mathbf{j}, \\
\mathbf{J} &= R\omega^3\sin(\omega t)\,\mathbf{i} - R\omega^3\cos(\omega t)\,\mathbf{j}
\end{aligned}$$

for the particle P. Also, we can write easily the following:

$$dx = -R\omega\sin(\omega t)\,dt,\quad dy = R\omega\cos(\omega t)\,dt,\quad dz = v_z dt,\quad ds = \sqrt{R^2\omega^2+{v_z}^2}dt.$$

So, the equalities

$$\frac{ds}{dt} = \sqrt{R^2\omega^2 + {v_z}^2}, \quad \frac{d^2s}{dt^2} = 0, \quad \frac{d^3s}{dt^3} = 0$$

hold.

As the path of the particle P, the right-handed circular helix on a cylinder of radius R is given below:

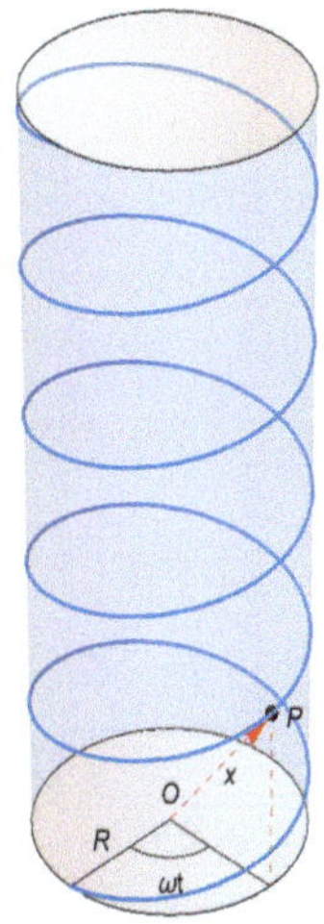

FIGURE 2.

It is easy to show that the helix can be parameterized by the arc-length $s = s(t) = t\sqrt{R^2\omega^2 + {v_z}^2}$ as

$$\gamma(s) = \left(R\cos\left(\frac{\omega s}{\sqrt{R^2\omega^2 + {v_z}^2}}\right),\ R\sin\left(\frac{\omega s}{\sqrt{R^2\omega^2 + {v_z}^2}}\right),\ \frac{v_z s}{\sqrt{R^2\omega^2 + {v_z}^2}}\right). \tag{4.1}$$

In that case, the Serret-Frenet basis for the helix can be calculated as follows:

$$\begin{aligned}
\mathbf{T} &= -\sin\alpha\sin\left(\frac{\omega s}{\sqrt{R^2\omega^2 + {v_z}^2}}\right)\mathbf{i} + \sin\alpha\cos\left(\frac{\omega s}{\sqrt{R^2\omega^2 + {v_z}^2}}\right)\mathbf{j} + \cos\alpha\mathbf{k},\\
\mathbf{N} &= -\cos\left(\frac{\omega s}{\sqrt{R^2\omega^2 + {v_z}^2}}\right)\mathbf{i} - \sin\left(\frac{\omega s}{\sqrt{R^2\omega^2 + {v_z}^2}}\right)\mathbf{j},\\
\mathbf{B} &= \cos\alpha\sin\left(\frac{\omega s}{\sqrt{R^2\omega^2 + {v_z}^2}}\right)\mathbf{i} - \cos\alpha\cos\left(\frac{\omega s}{\sqrt{R^2\omega^2 + {v_z}^2}}\right)\mathbf{j} + \sin\alpha\mathbf{k}.
\end{aligned} \tag{4.2}$$

It is not difficult to see that the curvature and the torsion are constant:

$$\kappa = \frac{R}{R^2 + ({v_z}^2/\omega^2)}, \quad \tau = \frac{v_z/\omega}{R^2 + ({v_z}^2/\omega^2)}.$$

By means of (2.4), the equalities

$$\begin{aligned}\tau_g &= \frac{v_z/\omega}{R^2+(v_z{}^2/\omega^2\,)} - \frac{d\varphi}{ds},\\ k_g &= \frac{R}{R^2+(v_z{}^2/\omega^2\,)}\cos\varphi,\\ k_n &= \frac{R}{R^2+(v_z{}^2/\omega^2\,)}\sin\varphi\end{aligned}$$

hold. By taking into consideration the relation (2.2) between the Serret-Frenet frame and Darboux frame and the equation (4.2), we obtain the first Darboux base as

$$\mathbf{T} = -\sin\alpha\sin\left(\frac{\omega s}{\sqrt{R^2\omega^2+v_z{}^2}}\right)\mathbf{i} + \sin\alpha\cos\left(\frac{\omega s}{\sqrt{R^2\omega^2+v_z{}^2}}\right)\mathbf{j} + \cos\alpha\,\mathbf{k}$$

second Darboux base as

$$\begin{aligned}\mathbf{Y} &= \left[-\cos\varphi\cos\left(\frac{\omega s}{\sqrt{R^2\omega^2+v_z{}^2}}\right) - \sin\varphi\cos\alpha\sin\left(\frac{\omega s}{\sqrt{R^2\omega^2+v_z{}^2}}\right)\right]\mathbf{i}\\ &\quad+\left[-\cos\varphi\sin\left(\frac{\omega s}{\sqrt{R^2\omega^2+v_z{}^2}}\right) + \sin\varphi\cos\alpha\cos\left(\frac{\omega s}{\sqrt{R^2\omega^2+v_z{}^2}}\right)\right]\mathbf{j}\\ &\quad+\left[-\sin\varphi\cos\alpha\right]\mathbf{k}\end{aligned}$$

and third Darboux base as

$$\begin{aligned}\mathbf{U} &= \left[-\sin\varphi\cos\left(\frac{\omega s}{\sqrt{R^2\omega^2+v_z{}^2}}\right) + \cos\varphi\cos\alpha\sin\left(\frac{\omega s}{\sqrt{R^2\omega^2+v_z{}^2}}\right)\right]\mathbf{i}\\ &\quad+\left[-\sin\varphi\sin\left(\frac{\omega s}{\sqrt{R^2\omega^2+v_z{}^2}}\right) - \cos\varphi\cos\alpha\cos\left(\frac{\omega s}{\sqrt{R^2\omega^2+v_z{}^2}}\right)\right]\mathbf{j}\\ &\quad+\left[\cos\varphi\sin\alpha\right]\mathbf{k}\end{aligned}$$

for the helix. From these last three equalities, (3.4) and (4.1),

$$q = \frac{s v_z\cos\alpha}{\sqrt{R^2\omega^2+v_z{}^2}} = t v_z\cos\alpha,\quad p = R,\quad b = \frac{s v_z\sin\alpha}{\sqrt{R^2\omega^2+v_z{}^2}} = t v_z\sin\alpha \tag{4.3}$$

can be written. By substituting (4.3) in (3.3), the equality

$$\mathbf{x} = (t v_z\cos\alpha)\,\mathbf{T} - (R\cos\varphi + t v_z\sin\alpha\sin\varphi)\,\mathbf{Y} + (-R\sin\varphi + t v_z\sin\alpha\cos\varphi)\,\mathbf{U}$$

is obtained. On the other hand,

$$r = \sqrt{R^2 + t^2\,v_z{}^2\cos^2\alpha}\,,\qquad r^* = t\,v_z$$

can be written by using (3.6) and (4.3).

Consequently, if Theorem 3.1 is applied, the components of the jerk are obtained as follows:

$$T_t = \frac{-R^2\omega^4 - \omega^2 {v_z}^2}{\sqrt{R^2\omega^2 + {v_z}^2}}, \quad T_r = 0, \quad T_{r^*} = \omega^2 v_z$$

for the particle P.

Acknowledgements

The authors would like to thank to Associate Professor Mahmut Akyiğit for his useful discussions, also thank to Dr. Furkan Semih Dündar for his help in drawing the Figure 2.

References

[1] F. Freudenstein, G. N. Sandor, *Mechanical design and systems hand-book*, Chapter 4, Page 13, McGraw-Hill Companies (1964).

[2] J. H. Bickford, *Mechanisms for intermittent motion*, Krieger Pub Co (1972).

[3] S. H. Schot, *Jerk: The time rate of change of acceleration*, American Journal of Physics **46** (1978) 1090–1094.

[4] P. Melchior, *Z. Ver. Dtsch. Ing.*, **72** (1928) 1842.

[5] H. Resal, *Traite de cinematique pure*, Mallet-Bachelier (1862).

[6] K. E. Özen, F. S. Dündar, M. Tosun, *An alternative approach to jerk in motion along a space curve with applications*, Journal of Theoretical and Applied Mechanics **57** (2019), 435–444.

[7] K. E. Özen, M. Güner, M. Tosun, *A Note on the acceleration and jerk in motion along a space curve*, Accepted for publication in An. St. Univ. Ovidius Constanta (2019).

[8] T. Shifrin, *Differential geometry: A first course in curves and surfaces*, University of Georgia (2015).

[9] B. O'Neill, *Elemantary differential geometry*, Academic Press, Newyork (1966).

[10] F. Doğan, Y. Yaylı, *Tubes with darboux frame*, Int. J. Contemp. Math. Sciences **7** (2012), 751–758.

[11] F. Doğan, *Generalized canal surfaces*, Ph.D. thesis, Ankara University Graduate School of Natural and Applied Sciences, Ankara (2012).

[12] Ö. Bektaş, S. Yüce, *Special smarandache curves according to Darboux frame in E^3*, Romanian Journal of Mathematics and Computer Science **3** (2013), 48–59.

[13] M. Tsirlin, *Jerk by axes in motion along a space curve*, Journal of Theoretical and Applied Mechanics **55** (2017), 1437–1441.

[14] K. E. Özen, M. Tosun, M. Akyiğit, *Siacci's theorem according to Darboux frame*, An. St. Univ. Ovidius Constanta-Seria Matematica **25** (2017), 155–165.

DOI 10.2478/9788395793882-008

Split octonion algebra and its zero divisors

Serpil Halici and Adnan Karataş*

Pamukkale University, Faculty of Arts and Sciences, Department of Mathematics, Denizli/TURKEY
*Corresponding Author

E-mail: shalici@pau.edu.tr, adnank@pau.edu.tr*

Abstract

Motivated by Miguel and Serodio's work [13] on the structure of quaternion rings over finite field, we studied some special elements of split octonion algebra over finite field, more precisely nilpotent and idempotent elements of split octonion algebra. For these special elements we gave some theorems and conditions which they should satisfy. Moreover, in this algebra we investigated the involutory elements and their properties.

2010 Mathematics Subject Classification. **17A20** 16N40
Keywords. octonions, idempotent elements, nilpotent elements.

1 Introduction

In this study, we examine idempotent and nilpotent elements in split octonion algebra. It is well-known that the idempotent and nilpotent elements are zero divisors. They are studied to find and give conditions for idempotent and nilpotent elements in degenerate quaternion algebras. In [15], the authors studied zero divisors especially idempotent and nilpotent elements of biquaternions. For the idempotent and nilpotent elements in split quaternion algebras over finite fields see [2,3,10]. In addition, the authors gave a survey about these special elements and they calculated the numbers of the elements in [13].

The pursue of characteristic features of idempotent and nilpotent elements can help coding theory, video processing, digital design, etc. (See [7, 12, 14, 17, 17]). In addition to idempotent and nilpotent elements, we introduce involutory elements which we have been inspired by involutory matrix [11]. Cryptography is one of the application areas of the involutory matrices [1].

The octonion algebra which is generalization of the quaternion algebra was defined by J. T. Graves in 1843 [4]. It is known that there are different constructions for octonion algebra. One of them is by using Cayley-Dickson process [9]. Cayley-Dickson process can be also used to obtain different algebras with isotropic or anisotropic norms which cause algebras to be split or real, respectively. Norm is isotropic if there exists a nonzero x with $\mathrm{N}(x) = 0$ else it is named as anisotropic norm [4,16].

Any element k of the octonion algebra is

$$k = a_0e_0 + a_1e_1 + \cdots + a_7e_7;\ \ a_0, a_1, \ldots, a_7 \in \mathbb{R}.$$

Conjugate and norm of the element k are

$$\overline{k} = a_0e_0 - a_1e_1 - \cdots - a_7e_7$$

and

$$\mathrm{N}(k) = \overline{k}k = k\overline{k} = a_0^2 + a_1^2 + \cdots + a_7^2$$

Received by the editors: 06 October 2019.
Accepted for publication: 24 December 2019 .

respectively.
After a brief summary of the octonion algebra, let us give some properties of split octonion algebras. In 1845, J. Cockle defined of the split quaternion algebra [6]. Then, his definition is used for different dimensional algebras. There are some important theorems in [16] about split algebras, one of them state *In each of the dimensions 2, 4 and 8 there is, up to isomorphism, exactly one split composition algebra.* Another important remark states that if the mentioned algebras defined over finite fields, then those algebras are split. Which means these algebras have non-trivial idempotent and nilpotent elements. In order to have better understanding of split octonion algebra and their subalgebras see [5]. In the first paper, the authors mentions null space which contains multiples of idempotent and nilpotent elements. The second paper is also mentions idempotent and nilpotent elements but on finite field $\mathbb{F}_{p^n}$ where p is a prime.
In the next section, we investigate split octonion algebra over finite fields which generalizes split quaternion algebra $\mathbb{H}/\mathbb{Z}_p$ and established conditions for idempotent and nilpotent elements. Also, we define involutory elements and give some properties of them.

2 Split octonion algebra and its zero divisors

As it is well-known, in any ring H, when $x^2 = x$ with $x \in H$ the element x is called as idempotent element. We know that the rings can be also characterized by idempotent element, for example a ring whose elements are all idempotent is called as Boolean ring which is also known as the idempotent ring.

Now, let us give the definition of split octonion algebra $\mathbb{O}/\mathbb{Z}_p$

$$\mathbb{O}/\mathbb{Z}_p = \{x|x = \sum_{i=0}^{7} a_i e_i; a_i \in \mathbb{Z}_p,\ p \text{ is a prime, } p \neq 2\}$$

where e_i's are basis elements of the octonion algebra $\mathbb{O}$.

Note that, due to the properties of octonions, split algebra $\mathbb{O}/\mathbb{Z}_p$ is neither commutative nor associative eight-dimensional algebra over $\mathbb{Z}_p$. Moreover, this algebra has zero divisors, nontrivial idempotent and nilpotent elements [16].

In the following theorem, we give some conditions for idempotent elements in $\mathbb{O}/\mathbb{Z}_p$.

Theorem 2.1. For the split algebra $\mathbb{O}/\mathbb{Z}_p$, following statements are true.

i) When the elements of $\mathbb{O}/\mathbb{Z}_p$ consist of only scalar part $x = a_0e_0$ then the idempotent elements of $\mathbb{O}/\mathbb{Z}_p$ are only 0 and 1.

ii) When the elements of $\mathbb{O}/\mathbb{Z}_p$ consist of scalar part and one imaginary unit such as $x = a_0e_0 + a_je_j$ with $j = 1, 2, \ldots, 7$ then the idempotent elements of $\mathbb{O}/\mathbb{Z}_p$ satisfy following equation

$$x = \frac{p+1}{2} + \sqrt{\frac{p^2-1}{4}} e_j.$$

iii) When the elements of $\mathbb{O}/\mathbb{Z}_p$ consist of scalar part and a set of imaginary units such as $x = a_0e_0 + V,\ V = \{a_ie_i \mid 1 \le i \le 3\}$ then the idempotent elements of $\mathbb{O}/\mathbb{Z}_p$ satisfy following equation

$$a_0 = \frac{p+1}{2} \text{ and } \mathrm{N}(V) = \frac{p^2-1}{4}.$$

iv) When the elements of $\mathbb{O}/\mathbb{Z}_p$ contain all of the basis units such as $x = a_0e_0 + a_1e_1 + \cdots + a_7e_7$ then the idempotent elements of $\mathbb{O}/\mathbb{Z}_p$ satisfy following equations

$$a_0 = \frac{p+1}{2} \text{ and } \mathrm{N}(a_1e_1 + a_2e_2 + \cdots + a_7e_7) = \frac{p^2-1}{4}.$$

Proof. i) Since $x = a_0e_0$ and $(x,p) = 1$ the claim is true.

ii) From the definition of idempotent element we can write

$$(a_0 + a_je_j)(a_0 + a_je_j) = a_0^2 - a_j^2 + 2a_0a_je_j \tag{2.1}$$

$$a_0 = a_0^2 - a_j^2, \tag{2.2}$$

$$a_j = 2a_0a_j. \tag{2.3}$$

By using equation (2.3), one can obtain $a_j = 0$ or $a_0 = \frac{p+1}{2}$. When $a_j = 0$, the idempotent element x is equal to 1 or 0. If $a_j \neq 0$, then a_j is calculated as follows.

$$a_0 = \frac{p+1}{2} \text{ and } \frac{p+1}{2} = \left(\frac{p+1}{2}\right)^2 - a_j^2$$

$$a_j = \sqrt{\frac{p^2-1}{4}}.$$

Hence, we showed the idempotent elements must satisfy $a_j = \sqrt{\frac{p^2-1}{4}}$. Now, in order to determine when $\frac{p^2-1}{4}$ is quadratic, we employ the Legendre Symbol

$$\left(\frac{\frac{p^2-1}{4}}{p}\right) = \left(\frac{p^2-1}{p}\right) = (p^2-1)^{\frac{p-1}{2}} = (-1)^{\frac{p-1}{2}} = \begin{cases} 1, & \text{if } p \equiv 1 \ (mod\ 4) \\ -1, & \text{if } p \equiv 3 \ (mod\ 4) \end{cases}. \tag{2.4}$$

So, if $p \equiv 1 \ (mod\ 4)$, then there are idempotent elements having the form $a_0e_0 + a_je_j$.

iii) The basis elements $\{e_1, e_2, e_3\}$ of V can be replaced with the triples $\{e_1, e_4, e_5\}$, $\{e_1, e_7, e_6\}$, $\{e_2, e_4, e_6\}$, $\{e_2, e_5, e_7\}$, $\{e_3, e_4, e_7\}$ and $\{e_3, e_6, e_5\}$. All these triples form subalgebras of the octonion algebra with $\{e_0\}$. These triples are isomorphic to each other and can be obtained by multiplication table or Fano plane (see [4, 8]). Similar to the prior part of the prove we can write following equations

$$(a_0 + V)(a_0 + V) = a_0^2 - Nr(V) + 2a_0V, \tag{2.5}$$

$$a_0 = a_0^2 - Nr(V) \tag{2.6}$$

and

$$a_i = 2a_0 a_i, \ \ 1 \le i \le 3. \tag{2.7}$$

In the equation (2.7), if we take $a_i = 0$, then $x = a_0$, else $a_0 = \frac{p+1}{2}$. Next, substituting $a_0 = \frac{p+1}{2}$ in the equation (2.6), we obtain following equation.

$$\mathrm{N}(V) = \frac{p^2-1}{4}.$$

Thus, the scalar and vectorial parts of idempotent elements are

$$a_0 = \frac{p+1}{2}$$

and

$$a_1^2 + a_2^2 + a_3^2 = \frac{p^2-1}{4},$$

respectively.

iv) If we consider W as $\{a_1e_1 + a_2e_2 + ... + a_7e_7\}$ and use the definition of idempotent element, then we obtain

$$a_0 = a_0^2 - \mathrm{N}(W), \tag{2.8}$$

$$a_i = 2a_0 a_i, \ \ 1 \le i \le 7. \tag{2.9}$$

For $1 \le i \le 7$, $a_i \neq 0$, the equation (2.9) means $a_0 = \frac{p+1}{2}$. Thus, $\mathrm{N}(W)$ is calculated as follows.

$$\mathrm{N}(W) = \frac{p^2-1}{4}.$$

Q.E.D.

Corollary 2.2. If x has no scalar part, then x is not an idempotent element of the $\mathbb{O}/\mathbb{Z}_p$.

Proof. Let x has at least one imaginary part and no scalar part. Then we calculate x^2. Because of the property of multiplication in $\mathbb{O}/\mathbb{Z}_p$ the result has to have a scalar part. From the hypothesis, x has no scalar part, any element without scalar part can not be idempotent element of $\mathbb{O}/\mathbb{Z}_p$. Q.E.D.

In the forthcoming theorems, we focus on the nilpotent elements of $\mathbb{O}/\mathbb{Z}_p$. First, let us recall the definition of the nilpotent element. For $y \in H$, if $y^k = 0, \ \ k \in \mathbb{N}$, then the element y is called as a nilpotent element of H.

Now, we give a lemma to investigate property of the nilpotent elements in $\mathbb{O}/\mathbb{Z}_p$.

Lemma 2.3. Let y be a nilpotent element of $\mathbb{O}/\mathbb{Z}_p$. The norm of y equals to zero.

Proof. Let y be any nilpotent element of the $\mathbb{O}/\mathbb{Z}_p$ with $y^k = 0$. From the properties of the octonion algebra $\mathbb{O}/\mathbb{Z}_p$ we have

$$y^2 + \mathrm{N}(y) - 2b_0 y = 0,$$
$$y^k(y - 2b_0)^k = (-\mathrm{N}(y))^k,$$
$$0 = (-\mathrm{N}(y))^k,$$

Thus, the norm of y is equal to zero as desired.

Q.E.D.

In the theorem 2, we introduce an additional attribute for the nilpotent elements.

Theorem 2.4. Let y be a nilpotent element of $\mathbb{O}/\mathbb{Z}_p$. Then y has no scalar part.

Proof. From Lemma 1, we conclude that $\mathrm{N}(y) = 0$, using the definition of nilpotent element and for least positive integer k which satisfies $y^k = 0$ we calculate the following equations.

$$y^2 - 2b_0 y = 0,$$
$$(y^2)^{(k+1)/2} = (2b_0)^{(k+1)/2} y^{(k+1)/2},$$

and

$$2b_0{}^{(k+1)/2} y^{(k+1)/2} = 0,$$

where k is odd. Analogously, for all even positive integers k, we get

$$y^2 - 2b_0 y = 0,$$
$$y^k = 2b_0{}^{(k/2)} y^{(k/2)}; \;\; y^k = 0,$$

and

$$2b_0{}^{(k/2)} y^{(k/2)} = 0, \;\; b_0 = 0.$$

Q.E.D.

Also, we can investigate general structure of the split octonion algebra $\mathbb{O}/\mathbb{Z}_p$ by introducing a new element and its properties. In order to accomplish this goal, we define the involutory element using the definition of involutory matrix [11].

In any ring H when $z^2 = 1_H$, the element z is called as an involutory element of H, where 1_H is identity element.

In the following theorem, we investigate the involutory elements of $\mathbb{O}/\mathbb{Z}_p$.

Theorem 2.5. For $z \in \mathbb{O}/\mathbb{Z}_p$, $z = c_0e_0 + \sum_{i=1}^{7} c_ie_i$, the following statements are satisfied.

i) The involutory elements which contain only scalar part such as $z = c_0e_0$ are only 1 and $p-1$.

ii) The involutory elements having the form $z = c_je_j$ where $j = 1, 2, \ldots, 7$, satisfy the following equality.

$$c_j = \sqrt{p-1}.$$

iii) The involutory elements having the form $z = \{c_1e_1 + c_2e_2 + c_3e_3\}$, have a norm which equals to $p-1$.

iv) For the involutory elements such as $z = \{c_1e_1 + c_2e_2 + ... + c_7e_7\}$, the norm must satisfy following equation

$$\mathrm{N}(z) = p-1.$$

Proof. i) Since $\mathbb{Z}_p$ is a field, the involutory elements are only 1 and $p-1$.

ii) Using the definition of involutory element, we get

$$-c_j^2 = 1 \tag{2.10}$$

From the equation (2.10), one can conclude that $c_j^2 = p-1$. Note that, if c_j not to be a square, then by using the equation (2.4) one can see that there are no involutory elements in the $\mathbb{O}/\mathbb{Z}_p$ having the form c_je_j, if $p \equiv 3 \ (mod\ 4)$.

iii) We take z as $z = \{c_1e_1 + c_2e_2 + c_3e_3\}$ and from the definition of involutory element we obtain that $\mathrm{N}(z) = -1$. Thus, the involutory elements which have the form $z = \{c_1e_1 + c_2e_2 + c_3e_3\}$ satisfy following equation;

$$c_1^2 + c_2^2 + c_3^2 = p-1.$$

iv) Analogously to the former proof, if we choose z as $z = \{c_1e_1 + c_2e_2 + ... + c_7e_7\}$, then we find that $\mathrm{N}(z) = -1$. Hence, components of the involutory elements satisfy following equation

$${c_1}^2 + {c_2}^2 + ... + {c_7}^2 = p-1.$$

Q.E.D.

Next corollaries are given without proofs.

Corollary 2.6. The involutory elements in $\mathbb{O}/\mathbb{Z}_p$ contain only one part, scalar or vectorial.

Corollary 2.7. An element is its own inverse if and only if it is involutory element in the $\mathbb{O}/\mathbb{Z}_p$.

Corollary 2.8. z is an involutory element in the $\mathbb{O}/\mathbb{Z}_p$ if and only if z satisfies the equation $(1+z)(1-z) = 0$.

3 Conclusion

In this study, we focus on the special elements of split octonion algebra over finite fields. We especially pay attention to some zero divisors of the algebra which are idempotent and nilpotent elements. Also, we define the involutory elements taking inspiration from involutory matrices. In a further study, one can calculate numbers of these elements in the $\mathbb{O}/\mathbb{Z}_p$.

References

[1] B. Acharya, G. S. Rath, S. K. Patra, and S. K. Panigrahy, *Novel methods of generating self-invertible matrix for hill cipher algorithm* (2007).

[2] M. Aristidou and A. Demetre, *Idempotent elements in quaternion rings over zp* International journal of Algebra, **6(27)** (2012) 249–254.

[3] M. Aristidou and A. Demetre, *A note on nilpotent elements in quaternion rings over zp*, International Journal of Algebra, **6(14)** (2012) 663–666.

[4] J. Baez, *The octonions*, Bulletin of the American Mathematical Society, **39(2)** (2002) 145–205.

[5] L. Bentz, *Subalgebras of the split octonions*, (2017).

[6] J. Cockle and T. S. Davies, *Lxiv. on certain functions resembling quaternions, and on a new imaginary in algebra*, Philosophical Magazine Series 3, **33(224)** (1848) 435–439.

[7] P. Di Giamberardino, S. Monaco, and D. Normand-Cyrot *Digital control through finite feedback discretizability*, Robotics and Automation, 1996. Proceedings., 1996 IEEE International Conference on, **4** (1996) 3141–3146.

[8] L. E. Dickson, *On quaternions and their generalization and the history of the eight square theorem*, Annals of Mathematics (1919) 155–171.

[9] C. Flaut, *Some equations in algebras obtained by the cayley-dickson process*, An. St. Univ. Ovidius Constanta **9(2)** (2001) 45–68.

[10] S. M. Kang, M. Munir, A. R. Nizami, M. Ali, and W. Nazeer, *On elements of split quaternions over zp*, Global Journal of Pure and Applied Mathematics **12(5)** (2016) 4253–4271.

[11] P. Lancaster and M. Tismenetsky, *The theory of matrices: with applications*, Elsevier (1985).

[12] T. Lin, *Achieving re-loss-free video coding*, IEEE Signal Processing Letters **16(4)** (2009) 323–326.

[13] C. Miguel and R. Serodio, *On the structure of quaternion rings over zp*, International Journal of Algebra **5(27)** (2011) 1313–1325.

[14] R. M. Murray, *Nilpotent bases for a class of nonintegrable distributions with applications to trajectory generation for nonholonomic systems*, Mathematics of Control, Signals and Systems **7(1)** (1994) 58–75.

[15] S. J. Sangwine and D. Alfsmann, *Determination of the biquaternion divisors of zero, including the idempotents and nilpotents*, Advances in applied Clifford algebras **20(2)** (2010) 401–410.

[16] T. A. Springer and F. D. Veldkamp, *Octonions, Jordan algebras and exceptional groups*, Springer (2013).

[17] Z. Zhu and T. Lin, *Idempotent h. 264 intraframe multi-generation coding*, Acoustics, Speech and Signal Processing, 2009. ICASSP 2009. IEEE International Conference on (2009) 1033–1036.

DOI 10.2478/9788395793882-009

Parallel Weingarten surface in Euclidean space with density e^z

Hande Kormalı[1] and Mahmut Akyiğit[2,*]

[1,2]Sakarya University, Department of Mathematics, Sakarya, Turkey
*Corresponding author
E-mail: hk.05@hotmail.com[1], makyigit@sakarya.edu.tr[2]

Abstract

In this study, the condition that the ruled surface of $\varphi(u,v)$ in Euclidean 3− space with density e^z is a Weingarten surface has been studied. While this review is being conducted, Q, F, J of $\varphi(u,v)$ are called structure functions of the ruled surface were used. Furthermore, the mean curvature H_ϕ and Gaussian curvature K_ϕ of the Weingarten surface were studied. Later on, the relationship between the mean H_ϕ^r curvature and Gauss curvature K_ϕ^r of this surface, which is the parallel surface $\varphi^r(u,v)$ of the Weingarten surface $\varphi(u,v)$ in Euclidean space with density e^z, was investigated. Weingarten surface $\varphi^r(u,v)$ condition of the parallel surface of the ruled Weingarten surface in Euclidean space with density e^z was examined.

2000 Mathematics Subject Classification. **14J26**. 53A05.
Keywords. parallel surface, space with density, Weingarten surface.

1 Introduction

The theory of surfaces occupies a wide place in various scientific fields such as physics, engineering, and architecture as well as mathematics. Apart from many architectures in daily life, mathematical models of various surfaces have been used in buildings including pyramids, temples, mosques, palaces, and tombs. Moreover, geometry has been also used in the interior and exterior decorations of these structures. Different types of surfaces have been used for many artifacts and materials. In the class of surfaces, one type is the developable surface that can be laid on the plane without being stretched or torn. Because of this feature, it is widely used in sheet-metal and plate-metal based industries, on the surface of a marine vessel, aircraft surfaces and architectural structures, [1]-[3]. Developable surfaces as a kind of ruled surfaces are classified into cylinders, cones or tangent surfaces of space curves, [4]. One of the important result related to the ruled surfaces is its relation with developable surfaces, although all developable surfaces are ruled ones, all ruled surfaces are not developable, [5, 6].

The ruled surfaces are very useful in many areas of sciences, for instance, Computer-Aided Manufacturing (CAM), Computer-Aided Geometric Desing (CAGD), geometric modelling and kinematics, [7]. It is possible to see many architectural structures based on ruled surfaces. For instance, at the end of the nineteenth century, the Shukhov Tower has been built by Vladimir Shukhov which is an example of the ruled surface.

The inner metric of a ruled surface determines the Gaussian curvature, therefore all the lengths and angles on the surface remain invariant in manufacturing. Hence, the ruled surfaces have been paid attention in engineering, architecture, and design, etc [8]-[11]. It is Gaspard Monge who pioneered the emergence of all these surfaces. Today, a great variety of studies have been devoted

Tbilisi Mathematical Journal Special Issue (IECMSA-2019), pp. 87–98.
Tbilisi Centre for Mathematical Sciences.

Received by the editors: 11 October 2019.
Accepted for publication: 10 April 2020.

to investigating the classifications of ruled surfaces. In [12], Izumiya classified the ruled surfaces into five classes such as developable surfaces, principal and binormal surfaces, Darboux developable surfaces, rectifying developable surfaces and focus surfaces of space curves.

Two parallel surfaces x and y have an identical distribution of normal vectors; i.e. their Gauss maps are indistinguishable. Thus, a family of parallel surfaces can be produced by translating a surface in the direction of its normal vectors by an equal amount everywhere on the surface.

If x and y are parallel surfaces, the formulae of the Gaussian and mean curvature of them yields the following interesting consequences: Firstly, the curvatures of surfaces parallel to minimal surfaces regarding a Bonnet transformation remain unamended. Secondly, increasing of the Gaussian curvature of a surface parallel to a minimal surface has been observed. This means that the minimal surface has a larger area than parallel surfaces, [13]. Park and Kim also showed that in 3−dimensional Euclidean space the parallel surfaces of non-developable ruled surfaces are not ruled surfaces, but the parallel surface of a developable ruled surface is a developable ruled surface [14].

Weingarten surfaces were introduced by Weingarten in 1861 in the context of the problem of finding all surfaces isometric to given surfaces of revolution, [15]. Weingarten improved his theory of $w-$surfaces in 1863, for this theory to be a Weingarten surface or $w-$surface of space in 3−dimensional Euclidean space, the principal curvature of the surface or the Gauss curvature K and the mean curvature H are described as linearly dependent with each other. The five classes of Weingarten surfaces are translation surfaces, channel surfaces of a curve which is a constant from its principal curvatures, helicoidal surfaces, constant mean curvature surfaces, and constant Gaussian curvature surfaces. Kuhnel took the Weingarten surface classification a step further by giving a condition as the dimensions of the structure functions $Q,\ F,\ J$ of the surface is constantly based on the condition of being a Weingarten surface [16]. In differential geometry, curves and space structures in density spaces are among the subjects studied recently. Morgan and Corwin studied the density spaces, [17, 18]. Inspired by these studies, the structures of different surfaces in different spaces of density have been started to be studied.

In this study, the definition of the ruled surface $\varphi(u,v)$ in 3−dimensional Euclidean space with density e^z has been introduced and also, the condition of being Weingarten has been examined. Some theorems and results were given by proving that the ruled surface $\varphi(u,v)$ is Weingarten. In 3−dimensional Euclidean space with density e^z, $\varphi(u,v)$ is a surface $\varphi^r(u,v)$ parallel to the ruled Weingarten surface. Then the condition of the surface being Weingarten parallel to the ruled Weingarten surface has been examined. In these regards, the theorems, corollaries and examples about $\varphi^r(u,v)$ according to the case of the ruled surface $\varphi(u,v)$ being developable and non-developable in 3−dimensional Euclidean space with density e^z.

2 Preliminaries

A ruled surface is a surface which can be swept out at least one 1-parameter family of lines in the space. For this reason, it has a parametrization of the form

$$\varphi(u,v) = c(u) + ve(u)$$

where c and e are called the base curve and director curve of the ruled surface, respectively. The foot of perpendicular bisector of two adjacent directrices of a ruled surface on the main directrix is called striction points. The geometric position of these points is also called the striction curve. The ruled surface is called non-developable ruled surface in case the drall, which is the ratio of the shortest distance between these two directrices to the angle between rectifiers, is non-zero. If $\varphi(u,v)$ is not a developable ruled surface such that $\langle e(u), e(u)\rangle = 1$ and $\langle e'(u), e'(u)\rangle = 1$, the directrix of $c(u)$ is a striction curve. Since $c(u)$ is the striction curve of $\varphi(u,v)$, $\langle c'(u), e'(u)\rangle = 0$, [19]. For brevity, the parameter u will not be written hereinafter.

Let $\{e,t,g\}$ be the spherical Frenet frame of spherical indicatrix vector of e. Then $t = e'$ and $g = e \times e'$. Derivative vectors for the spherical Frenet frame $\{e,t,g\}$ are given as follows:

$$\begin{aligned} e' &= t \\ t' &= -e - Jg \\ g' &= Jt \end{aligned}$$

where the vectors are said to be the central normal t and the asymptotic normal g of $\varphi(u,v)$, respectively. Moreover, here $J = \langle e'', e' \times e\rangle$ denotes the geodesic curvature κ_g of a spherical indicatrix curve e. On the other hand, the derivative of the striction curve c is given by

$$c' = Fe + Qg$$

where $F = \langle c', e\rangle$ and $Q = \langle c', e \times e'\rangle$.

The functions Q, F and J are called structure functions of a non-developable ruled surface $\varphi(u,v)$ in E^3. The structure functions F and Q of $\varphi(u,v)$ satisfy $F^2 + Q^2 = 1$ because the parameter u is also the arc-length parameter of the striction curve c of $\varphi(u,v)$, [19]. The Gaussian and mean curvatures of the ruled surface $\varphi(u,v)$ are

$$K = -\frac{Q^2}{D^4} \tag{2.1}$$

$$H = \frac{1}{2D^2}\left(Jv^2 - Q'v + Q(QJ - F)\right) \tag{2.2}$$

in terms of structure functions. By of the fact that $\varphi(u,v)$ is considered a non-developable ruled surface, K is non-zero. In that case, the function Q is non-zero everywhere, [19].

Suppose that M_1 and M_2 are two hypersurfaces in E^n and unit normal vector area of M_1 is

$$N_1 = \sum_{i=1}^{n} a_i \frac{\partial}{\partial x_i}, \quad a_i \in C^\infty(M, R).$$

M_2 is called the parallel hypersurface of M_1, if the function

$$\begin{aligned} f : M_1 &\to \quad M_2 \\ p &\to f(p) = (p_1 + ra_1(p), \ \dots, p_n + ra_n(p)) \end{aligned}$$

exists for a constant $r \in R$, [20]. After that, the surface parallel to the hypersurface M will be expressed as M^r.

Theorem 2.1. Let K and H denotes the Gaussian and mean curvatures of the surface of M at the point $p \in M$, respectively, and M^r denotes a parallel surface of the surface $M \subset E^3$. Then, the Gaussian and mean curvatures of the surface M^r at the point $f(P) \in M^r$ are given by

$$K^r = \frac{K}{1 + 2rH + r^2K} \tag{2.3}$$

and

$$H^r = \frac{H + rK}{1 + 2rH + r^2 K}, \tag{2.4}$$

respectively.

Under the conditions of the above theorem, the following equalities can be written

$$K = \frac{K^r}{1 - 2rH^r + r^2 K^r}$$

$$H = \frac{H^r - rK^r}{1 - 2rH^r + r^2 K^r}.$$

If the Gaussian and mean curvatures that given by the equations (2.3) and (2.4) of the parallel surface are rearranged in terms of structure functions,

$$K^r = \frac{-Q^2}{2D^4 + 4Dr\left(Jv^2 - Q'v + Q^2 J - QF\right) - 2r^2 Q^2}$$

$$H^r = \frac{D\left(Jv^2 - Q'v + Q^2 J - QF\right) - 2rQ^2}{2D^4 + 2Dr\left(Jv^2 - Q'v + Q^2 J - QF\right) - 2r^2 Q^2}$$

are obtained. Being a Weingarten surface in E^3 means that the changes of Gaussian and mean curvatures belonging to this surface are linearly independent of each other. That is

$$\Phi\left(K,\, H\right) = 0,$$

[21].

Theorem 2.2. Let $M\left(u,v\right) \subset E^3$ be a surface and K denotes the Gaussian curvature and H denotes the mean curvature of M. If

$$K_u H_v - K_v H_u = 0$$

holds for this surface, then M is a Weingarten surface, [21].

A manifold with a positive density function ϕ used to weight the volume and the hypersurface area. In terms of the underlying Riemannian volume dV_0 and area dA_0, the new, weighted volume and area are given by $dV = \phi dV_0$ and $dA = \phi dA_0$, respectively. For more details on manifolds with density, see [17, 18], [22, 23]. One of the best examples of density surfaces is the two-dimensional Gaussian plane. The Gaussian plane is the Euclidean plane with volume and length weighted by $\left(2\pi\right)^{-1} e^{-r^2/2}$. We can generalize the curvature of a curve or the mean curvature of a surface to manifolds with density.

Let $\phi\left(x\right)$ be the linear function in the form $\phi\left(x\right) = \sum\limits_{i=1}^{n} a_i x_i$ in Euclidean space E^n. Then $e^{\phi(x)}$ is called the log-linear density. We can write the density in e^{x_n} form by selecting the appropriate coordinates. Thus, we can examine the space E^n with density $e^{\phi(x)}$ while $E^{n-1} \oplus E_\phi$ is producing $\left(n-1\right)-$dimensional Euclidean space E^{n-1} and E_ϕ to real lines with density e^{x_n}.

Taking into account that $\nabla\phi = (0,0,...,1)$, it is seen that the following equation exists;

$$\frac{d\phi}{dN} = \langle \nabla\phi, N\rangle = \|\nabla\phi\| . \|N\| . \cos\theta\,(\nabla\phi, N) = \cos\theta\,(\nabla\phi, N).$$

Here $\frac{d\phi}{dN} = \langle \nabla\phi, N\rangle$ is the cosine of the angle between N and z-axis [24]. Riemann curvature with density is defined by

$$\kappa_\phi = \kappa - \frac{d\phi}{dN}$$

where κ is Riemann curvature, [17]. Besides, for the case $n = 3$, we can give the Gaussian and mean curvatures with density,

$$K_\phi = K - \Delta\phi \tag{2.5}$$

and

$$H_\phi = H - \frac{1}{2}\frac{d\phi}{dN}, \tag{2.6}$$

respectively [17],[24]. These curvatures are also called Gaussian $\phi-$curvature and mean $\phi-$curvature. where K and H are Gaussian and mean curvatures of the ruled surface, respectively.

In view of the fact that $\nabla\phi = (0,0,1)$ and $\Delta\phi = 0$, Gauss and mean $\phi-$curvatures of the ruled surface in terms of structure functions are obtained as follows

$$K_\phi = -\frac{Q^2}{D^4} \tag{2.7}$$

$$H_\phi = \frac{1}{2D^3}\left(Jv^2 - Q'v + Q\,(QJ - F)\right) - \frac{1}{2D}\langle \nabla\phi, Qt - vg\rangle \tag{2.8}$$

with the aid of equations (2.5) and (2.6) in E^3 with density e^z where $D = \sqrt{Q^2 + v^2}$.

3 Ruled surfaces in E^3 with density e^z

In this section, the ruled Weingarten surfaces in 3−dimensional Euclidean space E^3 with density e^z and parallel surfaces of these surfaces will be examined.

3.1 Ruled Weingarten surfaces in Eculidean space with density e^z

First, we investigate the state of a ruled surface to be Weingarten by using the Gaussian and mean $\phi-$ curvatures of the ruled surface $\varphi\,(u,v)$ in the Euclidean space with density.

If there exists the Jacobi equation $\Phi\,(H_\phi, K_\phi) = 0$ between the Gaussian and the mean $\phi-$curvatures in the Euclidean space with density e^z, it is called Weingarten surface.

Theorem 3.1. Let K_ϕ and H_ϕ denote the Gaussian and mean $\phi-$curvatures of the ruled surface $\varphi\,(u,v)$ in the 3−dimensional Euclidean space with density e^z, respectively. If the following equality

$$(K_\phi)_u (H_\phi)_v - (K_\phi)_v (H_\phi)_u = 0 \tag{3.1}$$

holds, the ruled surface $\varphi\,(u,v)$ is the Weingarten surface.

Proof. Let us first observe the derivatives of H_ϕ and K_ϕ with respect to parameters u and v. The derivatives of the equation (2.7) with respect to parameters u and v are

$$(K_\phi)_u = \frac{2QQ'\left(Q^2+v^2\right)}{D^6} \quad , \quad (K_\phi)_v = \frac{4vQ^2}{D^6} \tag{3.2}$$

respectively. The central and asymptotic normals of the ruled surface $\varphi(u,v)$ are vectors $t = (t_1, t_2, t_3)$ and $g = (g_1, g_2, g_3)$, respectively. If one takes the derivative of the equation (2.6) with respect to u and v by considering t, g and $\nabla\phi = (0,0,1)$, the following have been obtained

$$(H_\phi)_u = \frac{S_u}{2D^5} \quad , \quad (H_\phi)_v = \frac{S_v}{2D^5}. \tag{3.3}$$

The equalities of the expression of S_u and S_v can be obtained by basic calculations, left to the reader.

Let us arrange the equation (3.1) that is provided by the equations (3.2) and (3.3). This gives

$$(K_\phi)_u(H_\phi)_v - (K_\phi)_v(H_\phi)_u = \frac{\left(2QQ'v^2 + 2Q'Q^3\right)S_v - 4vQ^2S_u}{2D^{11}}.$$

So that this surface is a Weingarten surface in the Euclidean space with density e^z, the right side of the equation must be zero. Since the Gaussian curvature of K given by the equation (2.1) is different from zero, $Q' = F' = J' = 0$ and ${t'}_3 = {g'}_3 = 0$ must be held in the neighbourhood of any point satisfying the condition $Q \neq 0$. In this case, the values Q, F, J, g_3 and t_3 must be constant. Moreover, $J = F = 0$ when $H = 0$. In this case, the surface is helicoidal surface. Q.E.D.

Hereunder this theory, we can reach the following result.

Result 3.2. For a non-developable $\varphi(u,v)$ ruled surface in the 3−dimensional Euclidean space with density e^z, the following conditions hold;

i) φ is a Weingarten surface.

ii) The values Q, F, J, g_3, t_3 are constants.

Example 3.3. Let φ be a ruled surface in 3−dimensional Euclidean space with density e^z defined as

$$\varphi(u,v) = \left(-\cos(u) + v\sin(u), -\sin(u) - v\cos(u), u\right).$$

The Gaussian and mean curvatures of the ruled surface $\varphi(u,v)$ are

$$K = -\frac{1}{\left(1+v^2\right)^2} \quad , \quad H = -\frac{1}{2\left(1+v^2\right)^{3/2}} \tag{3.4}$$

respectively. This non-developable helicoidal surface in Euclidean space is a ruled Weingarten surface. Let us show that this surface is a Weingarten surface in Euclidean space E^3 with density e^z.

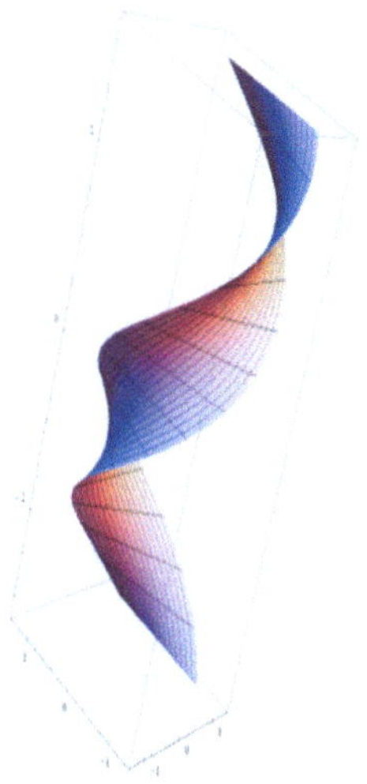

FIGURE 1. Non-developable ruled surfaces $\varphi(u,v)$

Let us substitute (3.4) in (2.7) and (2.8) such that $\nabla\phi=(0,0,1)$. So, the Gaussian $\phi-$curvature and mean $\phi-$curvature of the ruled surface $\varphi(u,v)$ in Euclidean space with density e^z are as follows:

$$K_\phi=-\frac{1}{(1+v^2)^2}\quad,\quad H_\phi=\frac{-1+v+v^3}{2(1+v^2)^{3/2}}.$$

If we calculate the partial derivative of the above equations according to the parameters u and v, and replace them in the equation (3.1), the equation

$$(K_\phi)_u(H_\phi)_v-(K_\phi)_v(H_\phi)_u=0$$

is obtained. So the non-developable ruled surface $\varphi(u,v)$ in 3−dimensional Euclidean space with density e^z is a Weingarten surface.

Result 3.4. Let $\varphi(u,v)$ be a ruled Weingarten surface in 3−dimensional Euclidean space. Then, the ruled surface $\varphi(u,v)$ is also Weingarten in the 3−dimensional Euclidean space with density e^z.

3.2 Parallel surfaces of ruled Weingarten surfaces in Euclidean space with density e^z

Let $\varphi^r(u,v)$ be the parallel surface of the ruled Weingarten surface $\varphi(u,v)$ in Euclidean space with density e^z. Let us examine whether the parallel surface is Weingarten surface using Gaussian $\phi-$curvature and mean $\phi-$curvature.

Let the coefficients of the first and the second fundamental forms of parallel surface $\varphi^r(u,v)$ in Euclidean space with density e^z be E^r_ϕ, F^r_ϕ, G^r_ϕ and L^r_ϕ, M^r_ϕ, N^r_ϕ. In this case, there are the following equations,

$$\begin{aligned}{E_\phi}^r&=E_\phi-2rL_\phi+r^2\langle N_u,N_u\rangle\\{F_\phi}^r&=F_\phi-2rM_\phi+r^2\langle N_u,N_v\rangle\\{G_\phi}^r&=G_\phi+r^2\langle N_v,N_v\rangle\end{aligned}$$

and

$$\begin{aligned} L_{\phi}{}^{r} &= L_{\phi} - r\left\langle N_u, N_u \right\rangle \\ M_{\phi}{}^{r} &= M_{\phi} - r\left\langle N_u, N_v \right\rangle \\ N_{\phi}{}^{r} &= -r\left\langle N_v, N_v \right\rangle . \end{aligned}$$

By considering $\nabla\phi = (0,0,1)$ and $\Delta\phi = 0$, the Gaussian $\phi-$curvature and mean $\phi-$curvature of the parallel surface $\varphi^r(u,v)$ in Euclidean space with density e^z, are

$$K_{\phi}^{r} = \frac{-Q^2}{2D^4 + 4Dr\left(Jv^2 - Q'v + Q^2J - QF\right) - 2r^2Q^2}, \tag{3.5}$$

and

$$H_{\phi}^{r} = \frac{D\left(Jv^2 - Q'v + Q^2J - QF\right) - 2rQ^2}{2D^4 + 2Dr\left(Jv^2 - Q'v + Q^2J - QF\right) - 2r^2Q^2} - \frac{Qt_3 - vg_3}{2D}, \tag{3.6}$$

respectively.

Theorem 3.5. Let $\varphi^r(u,v)$ be the parallel surface of a non-developable ruled surface $\varphi(u,v) = \alpha(u) + vX(u)$ in the Euclidean space with density e^z. If the equations

$$\|X\| = \|X'\| = 1\ , \quad \langle \alpha', X' \rangle = 0, \quad K \neq 0,$$
$$F = \langle \alpha', X \rangle\ , \quad Q = \det(\alpha', X, X')\ , \quad J = \det(X'', X', X)$$

exist, the following conditions are equivalent. If $\varphi(u,v)$ is a Weingarten surface, then $\varphi^r(u,v)$ is a Weingarten surface.

Proof. Let $\varphi(u,v)$ is a Weingarten surface. In this condition, the values of Q, F, J, t_3 and g_3 are constants. If partial derivatives of the Gaussian $\phi-$curvature of the parallel surface given by the equation (3.5) are taken according to parameters u and v, the following equations are obtained

$$\left(K_{\phi}^{r}\right)_u = \frac{\psi_u}{D\left(Q^4 + v^4 + 2Q^2v^2 + Dr\left(Jv^2 - Q'v + Q^2J - QF\right) - 2Q^2r^2\right)} \tag{3.7}$$

and

$$\left(K_{\phi}^{r}\right)_v = \frac{\psi_v}{D\left(Q^4 + v^4 + 2Q^2v^2 + Dr\left(Jv^2 - Q'v + Q^2J - QF\right) - 2Q^2r^2\right)}, \tag{3.8}$$

respectively. The equalities of the expression of ψ_u and ψ_v can be obtained by basic calculations, left to the reader.

If the partial derivatives of the mean $\phi-$curvature of the parallel surface given by the equation (3.6) are taken with respect to the parameters u and v, the following equations are obtained

$$\left(H_{\phi}^{r}\right)_u = \frac{\mu_u}{D(2D^4 + 2Dr\left(Jv^2 - Q'v + Q^2J - QF\right) - 2Q^2r)^2} - \frac{\varpi_u}{4D^3} \tag{3.9}$$

and

$$\left(H_{\phi}^{r}\right)_v = \frac{\mu_v}{D(2D^4 + 2Dr\left(Jv^2 - Q'v + Q^2J - QF\right) - 2Q^2r)^2} + \frac{Dg_3}{v}, \tag{3.10}$$

respectively. The equalities of the expression of ϖ_u, μ_u and μ_v can be obtained by basic calculations, left to the reader. If the values found in equations (3.7)-(3.10) are substituted in the equation $\left(K_\phi^r\right)_u\left(H_\phi^r\right)_v - \left(K_\phi^r\right)_v\left(H_\phi^r\right)_u$ and the conditions of ii) in the Result 3.2 are taken into account, we have the following equation

$$\left(K_\phi^r\right)_u\left(H_\phi^r\right)_v - \left(K_\phi^r\right)_v\left(H_\phi^r\right)_u = 0.$$

This completes the proof.

Q.E.D.

Theorem 3.6. Let $\varphi^r(u,v)$ be the parallel surface of the developable ruled Weingarten surface $\varphi(u,v)$ in the Euclidean space with density e^z. Then $\varphi^r(u,v)$ is a developable ruled Weingarten surface.

Proof. Let $\varphi(u,v) = \alpha(u) + vX(u)$ be a developable ruled Weingarten surface in Euclidean space with density e^z. The normal of the developable ruled surface is

$$N_1 = \alpha_u(u) \wedge X(u) + v\left(X_u(u) \wedge X(u)\right).$$

Since the vectors $\vec{\alpha}_u(u) \wedge X(u)$ and $X_u(u) \wedge X(u)$ are linearly dependent, then

$$\alpha_u(u) \wedge X(u) = \varepsilon\left(X_u(u) \wedge X(u)\right) \tag{3.11}$$

can be written, [14].

If one substitutes the equation (3.11) in the equation of N_1, the following equality is obtained:

$$N_1 = (\varepsilon + v)\left(X_u \wedge X\right).$$

Let the unit normal of a surface in Euclidean space with density e^z be N, we found that

$$N = \frac{N_1}{\|N_1\|} = X_u \wedge X.$$

The parallel surface of developable Weingarten surface $\varphi(u,v)$ in Euclidean space with density e^z is

$$\varphi^r(u,v) = \alpha(u) + r\left(X_u(u) \wedge X(u)\right) + vX(u)$$

and the drall of $\varphi^r(u,v)$ is

$$\lambda = \langle \alpha_u, X_u \wedge X\rangle + r\langle X_{uu} \wedge X, X_u \wedge X\rangle.$$

From these calculations, it is seen that λ is zero. So $\varphi^r(u,v)$ is the developable ruled Weingarten surface in Euclidean space with density e^z. Thus the proof is completed. Q.E.D.

Theorem 3.7. The principal curvatures $k_{1\phi}$ and $k_{2\phi}$ of a surface $\varphi(u,v)$ in E^3 with density e^z are the roots of the following quadratic equation:

$$\lambda^2 - \lambda\left(2H_\phi - \frac{d\phi}{dN}\right) + \left(K_\phi - 2H_\phi\frac{d\phi}{dN} + \nabla\phi\right) = 0.$$

Here the principal curvatures of $k_{1\phi}$ and $k_{2\phi}$ are obtained as follows:

$$k_{1\phi}=\frac{\left(2H_\phi-\frac{d\phi}{dN}\right)+\sqrt{4H_\phi{}^2+\left(\frac{d\phi}{dN}\right)^2-4K_\phi+4H_\phi\frac{d\phi}{dN}-4\nabla\phi}}{2}$$

$$k_{2\phi}=\frac{\left(2H_\phi-\frac{d\phi}{dN}\right)-\sqrt{4H_\phi{}^2+\left(\frac{d\phi}{dN}\right)^2-4K_\phi+4H_\phi\frac{d\phi}{dN}-4\nabla\phi}}{2}.$$

Proof. Let $\varphi(u,v)$ be a ruled surface in Euclidean space with the density of e^z and $k_{1\phi}$, $k_{2\phi}$ denote the principal curvatures of this surface. The equation $\det(A-\lambda\mathrm{I})=0$ provides us a characteristic equation such that $A=\begin{pmatrix}k_{1\phi} & 0\\ 0 & k_{2\phi}\end{pmatrix}$ is a matrix and $\lambda$is a constant.

To find the roots of this equation, let's calculate Gaussian $\phi-$curvature and mean $\phi-$curvature. The Gaussian ϕ - curvature in terms of principle curvatures $k_{1\phi}$ and $k_{2\phi}$ is

$$\begin{aligned}K_\phi=k_1k_2-\nabla\phi&=\left(k_{1\phi}+\tfrac{d\phi}{dN}\right)\left(k_{2\phi}+\tfrac{d\phi}{dN}\right)-\nabla\phi\\&=k_{1\phi}k_{2\phi}+\tfrac{d\phi}{dN}\left(2H_\phi-\tfrac{d\phi}{dN}\right)+\left(\tfrac{d\phi}{dN}\right)^2-\nabla\phi\end{aligned}$$

and then the product of roots is

$$k_{1\phi}k_{2\phi}=K_\phi-2H_\phi\frac{d\phi}{dN}+\nabla\phi.$$

The mean $\phi-$curvature in terms of principle curvatures $k_{1\phi}$ and $k_{2\phi}$ is

$$\begin{aligned}H_\phi=\frac{k_1+k_2-\frac{d\phi}{dN}}{2}&=\frac{\left(k_{1\phi}+\frac{d\phi}{dN}\right)+\left(k_{2\phi}+\frac{d\phi}{dN}\right)-\frac{d\phi}{dN}}{2}\\&=\tfrac{1}{2}\left(k_{1\phi}+k_{2\phi}+\tfrac{d\phi}{dN}\right)\end{aligned}$$

and then the sum of roots is found as

$$k_{1\phi}+k_{2\phi}=2H_\phi-\frac{d\phi}{dN}.$$

By considering the relations for $k_{1\phi}k_{2\phi}$ and $k_{1\phi}+k_{2\phi}$ the equation $\det(A-\lambda\mathrm{I})=0$ give us

$$\begin{aligned}\mathrm{P}_s(\lambda)&=\det\left(\begin{pmatrix}k_{1\phi} & 0\\ 0 & k_{2\phi}\end{pmatrix}-\lambda.\begin{pmatrix}1 & 0\\ 0 & 1\end{pmatrix}\right)\\&=k_{1\phi}k_{2\phi}-\lambda\left(k_{1\phi}+k_{2\phi}\right)+\lambda^2\\&=\lambda^2-\lambda\left(2H_\phi-\tfrac{d\phi}{dN}\right)+\left(K_\phi-2H_\phi\tfrac{d\phi}{dN}+\nabla\phi\right)=0.\end{aligned}$$

The roots of this quadratic equation are

$$k_{1\phi}=\frac{\left(2H_\phi-\frac{d\phi}{dN}\right)+\sqrt{4H_\phi{}^2+\left(\frac{d\phi}{dN}\right)^2-4K_\phi+4H_\phi\frac{d\phi}{dN}-4\nabla\phi}}{2}$$

and

$$k_{2\phi}=\frac{\left(2H_\phi-\frac{d\phi}{dN}\right)-\sqrt{4H_\phi{}^2+\left(\frac{d\phi}{dN}\right)^2-4K_\phi+4H_\phi\frac{d\phi}{dN}-4\nabla\phi}}{2},$$

respectively.

Q.E.D.

References

[1] J. Boersma and J. Molenaar, *Geometry of the shoulder of a packaging machine*, SIAM Review **37**(3) (1995) 406-422.

[2] W. H. Frey and M. J. Mancewicz, *Develepable surfaces: Properties, representations and methods of desing*, Technical Report, GM Research Publication (1992) GMR-7637.

[3] J. Pegna and E-E. Wolter, *Geometrical criteria to guarantee curvature contiuity of blend surfaces*, ASME Journal of Mechanical Desing **114** (1992) 201-210.

[4] R. Abdel-Baky and Y. Ünlütürk, *The relatively osculating developable surfaces of a surface along a direction curve*, Communications Faculty of Sciences University of Ankara Series A1 **69** (2020) 511-527.

[5] S. Lawrence, *Developable surfaces: Their history and application*, Nexus Netw. J. **13** 701-714.

[6] V. Ushakow, *Developable surfaces in Euclidean space*, Journal of the Australian Mathematical Society Series A **66** (1999) 388-402.

[7] G. Y. Şentürk and S. Yüce, *On the evolute offsets of ruled surfaces using the Darboux frame*, Communications Faculty of Sciences University of Ankara Series A1-Mathematics and Statistics **68**(2) (2019) 1256-1264.

[8] T. Sasaki, *Projektive differantial geometry and linear homogeneus differantial equations*, Rokko Lectures in Mathematics, Kobe University, **5** (1999).

[9] S. Aydöner and K. Arslan, *Erratum to "On Convolution surfaces in Euclidean spaces"* J. Math. Sci. Model. *1*(2) (2018), 86-92, Journal of Mathematical Sciences and Modelling *2*(1) (2019) 71-73.

[10] M. Schneider, *Interpolation with developable strip-surfaces consisting of cylinders and cones*, in Mathematical Methods for curves and surfaces II, ed. by M. Daehlen, T. Lyche and L.L. Schumacker, Vanderbilt University Press (1998) 437-444.

[11] I. I. Papadopoulou and S. Stamatakis, *On polar relative normalizations of ruled surfaces*, Universal Journal of Mathematics and Applications **1**(2) (2018) 74-79.

[12] S. Izumiya, K. Sji and N. Takecuni, *Circular surfaces*, Advances in Geometry **7** (2005) 295–313.

[13] S. Hyde and S. Lıdın, *The Mathematics of curvature*, The role of curvature in condensed matter: Physics, Chemistry and Biology (1997) 1–42.

[14] K. R. Park and G. I. Kim, *Offsets of ruled surfaces*, Journal of Korean Computer Graphics Society, **4** (1998) 69–75.

[15] Yu Fu and Lan Li, *A class of Weingarten surfaces in Euclidean 3-space*, (2013) www.hindawi.com/journals/aaa/2013/398158

[16] W. Kühnel, *Ruled W-surfaces*, Archiv der Mathematik **62** (1994) 475–480.

[17] I. Corwin, N. Hoffman, S. Hurder, V. Sesum and Y. Xu, *Differential geometry of manifolds with density*, Rose-Hulman Und. Math. J. **7(1)** (2006) 1–15.

[18] F. Morgan, *Manifolds with density*, Notices of the American Mathematical Society **52**(8) (2005) 853-858.

[19] D. W. Yoon, *On the offsets of ruled surfaces in Euclidean space*, International Journal of Pure and Applied Mathematics **108(4)** (2016) 985–997.

[20] H. H. Hacısalihoğlu, *Diferansiyel geometri*, Ankara Üniversitesi, Fen Fakültesi, Matematik Bölümü (1994).

[21] Z. M. Sipus, *Ruled Weingarten surfaces in Galilean space*, Periodica Mathematica Hungarica **56**(2) (2008) 213–225.

[22] L. Belarbi, *Some results in Riemannian manifolds with density*, Analele Univ. Oradea Fasc. Matematica, **XXII**(2) (2015) 81–86.

[23] Ö. G. Yıldız, *Constructions of helicoidal surfaces in a 3-dimensional complete manifold with density*, Mathematics **7**, 27 (2019).

[24] D. T. Hieu and N. M. Hoang, *Ruled minimal surfaces in R^3 with density e^z*, Pasific Journal of Mathematics **243**(2) (2009) 277–285.

DOI 10.2478/9788395793882-010

Stability inverse problem for diffusion operators

Etibar S. Panakhov[1] and Ahu Ercan[2]

[1]Baku State University, Institute of Applied Mathematics, Baku, Azerbaijan
[2] Firat University, Department of Mathematics, Elazig, Turkey
E-mail: epenahov@hotmail.com[1], ahuduman24@gmail.com[2]

Abstract

In this study, the certain stability problem is studied for diffusion equation by special method on a finite interval. The method which used was given firstly by Ryabushko for Sturm-Liouville problem in [23]. There are many different stability criteria. In practice, any one of a number of different stability criteria is applied. The stability of spectral functions has been shown by using the asymptotics of the eigenvalues for diffusion problems with different initial conditions in this work. Also, we introduce the notion of norming constants and establish their interrelation with the spectra.

2010 Mathematics Subject Classification. **34B09** 34D20. 34L05
Keywords. Stability, spectral function, diffusion equation.

1 Introduction

Inverse problems of spectral analysis are the problems of constructing a linear operator from some of its spectral characteristics. An effective method of constructing Sturm-Liouville operator from two spectra is given in [3]. Levitan [4] established determination of the Sturm-Liouville operator from one and two spectra. Jaulent and Jean [6] stated the actual background of diffusion operators and discussed the inverse problem for the diffusion operator. Gasymov and Guseinov [1] studied the spectral theory of diffusion operators. Panakhov and Koyunbakan [12] considered the half inverse problem for the diffusion operators on the interval. Sat and Panakhov studied generalized degeneracy of the kernels for the diffusion operator in [14]. Direct and inverse problems for the diffusion operator were considered in several works [6]- [14]. In this paper, we shall establish certain stability of spectral functions of two spectral problems for diffusion operators on a finite interval. The stability problem means that when the eigenvalues of the considered problems finitely coincide then the bound of variation for the spectral functions is estimated. In study [23], the author showed the proximity of the spectral functions for regular Sturm-Liouville problems on a finite interval. This type of problem for a self-adjoint operator is considered by Marchenko [18] on a semi-axis. There are many researches about the stability of the spectral theory in the literature [15]- [25]. In this work, the coefficients $p(x)$ and $q(x)$ of quadratic pencils of Sturm-Liouville operators are uniquely determined by two spectra using classic way for Sturm-Liouville operators [2, 5]. The stability problem for diffusion equation is discussed on finite interval using the method given in study [23]. We show the proximity of the spectral functions for diffusion equations.

2 Preliminaries

Consider the diffusion equation

$$-y''(x) + (2\lambda p(x) + q_1(x))\, y(x) = \lambda^2 y(x) \quad, 0 \le x \le \pi \tag{2.1}$$

Tbilisi Mathematical Journal Special Issue (IECMSA-2019), pp. 99–109.
Tbilisi Centre for Mathematical Sciences.

Received by the editors: 25 February 2020.
Accepted for publication: 10 April 2020.

with the boundary conditions

$$y'(0) - h_1 y(0) = 0, \quad y'(\pi) + Hy(\pi) = 0 \tag{2.2}$$

$$y'(0) - h_2 y(0) = 0, \quad y'(\pi) + Hy(\pi) = 0 \tag{2.3}$$

where λ is a spectral parameter, $p(x) \in W_2^1[0,\pi]$ and $q_1(x) \in L_2[0,\pi]$, $p(x)$ and $q_1(x)$ are real-valued functions, h_1, h_2 and H are real numbers with $h_1 \neq h_2$. Let us denote the eigenvalues of the boundary value problems (2.1)-(2.2) and (2.1), (2.3) by $\{\lambda_{1,n}\}_{-\infty}^{\infty}$ and $\{\mu_{1,n}\}_{-\infty}^{\infty}$, respectively. The asymptotic expressions for the eigenvalues are given by [1]

$$\lambda_{1,n} = n + c_0 + \frac{c_1}{n} + \frac{c_{1,n}}{n}, \tag{2.4}$$

$$\mu_{1,n} = n + c_0 + \frac{\tilde{c}_1}{n} + \frac{\tilde{c}_{1,n}}{n}, \tag{2.5}$$

for large negative and positive values of n, respectively, where

$$c_0 = \frac{1}{\pi}\int_0^\pi p(x)\,dx, \quad \sum_n |c_{1,n}|^2 < \infty, \quad \sum_n |\tilde{c}_{1,n}|^2 < \infty,$$

$$c_1 = \frac{1}{\pi}\left(h_1 + H + \frac{1}{2}\int_0^\pi \left(q_1(x) + p^2(x)\right)dx\right),$$

$$\tilde{c}_1 = \frac{1}{\pi}\left(h_2 + H + \frac{1}{2}\int_0^\pi \left(q_1(x) + p^2(x)\right)dx\right).$$

Now consider the second equation

$$-y''(x) + (2\lambda p(x) + q_2(x))\,y(x) = \lambda^2 y(x) \quad, 0 \le x \le \pi \tag{2.6}$$

with the boundary conditions (2.2) and (2.3), where $q_2(x)$ is real-valued function and $q_2(x) \in L_2[0,\pi]$. Denote by$\{\lambda_{2,n}\}_{-\infty}^{\infty}$ and $\{\mu_{2,n}\}_{-\infty}^{\infty}$ the eigenvalues of problems (2.2), (2.6), and (2.3), (2.6), respectively, and the following asymptotic formulas are hold [1]

$$\lambda_{2,n} = n + c_0 + \frac{c_2}{n} + \frac{c_{2,n}}{n}, \tag{2.7}$$

$$\mu_{2,n} = n + c_0 + \frac{\tilde{c}_2}{n} + \frac{\tilde{c}_{2,n}}{n}, \tag{2.8}$$

where

$$\sum_n |c_{2,n}|^2 < \infty, \quad \sum_n |\tilde{c}_{2,n}|^2 < \infty,$$

$$c_2 = \frac{1}{\pi}\left(h_1 + H + \frac{1}{2}\int_0^\pi \left(q_2(x) + p^2(x)\right)dx\right),$$

$$\tilde{c}_2 = \frac{1}{\pi}\left(h_2 + H + \frac{1}{2}\int_0^\pi \left(q_2(x) + p^2(x)\right)dx\right).$$

$\varphi(x,\lambda)$ is the solution of the equation (2.1) satisfying initial conditions

$$\varphi(0,\lambda)=1\ ,\ \ \varphi'(0,\lambda)=h_1. \tag{2.9}$$

Obviously, $\varphi_n(x)=\varphi(x,\lambda_n)$ is an eigenfunction of (2.1)-(2.2), corresponding to the eigenvalue λ_n and the solution of the problem (2.1)-(2.2) is given by [1]

$$\varphi(x,\lambda)=\cos(\lambda x-\alpha_{h_1}(x))+\int_0^x A_{h_1}(x,t)\cos\lambda t dt+\int_0^x B_{h_1}(x,t)\sin\lambda t dt$$

where

$$\alpha_{h_1}(x)=xp(0)+2\int_0^x\left(A_{h_1}(\xi,\xi)\sin\alpha(\xi)-B_{h_1}(\xi,\xi)\cos\alpha(\xi)\right)d\xi.$$

Let's set the norming constants of problem (2.1)-(2.2)

$$\alpha_{1,n}=\int_0^\pi\varphi_n^2(x)\,dx-\frac{1}{\lambda_n}\int_0^\pi p(x)\,\varphi_n^2(x)\,dx.$$

Similarly, $\zeta(x,\lambda)$ is the solution of the equation (2.6) satisfying conditions (2.9).
Then denote the norming constants of problem (2.6), (2.2) as follow:

$$\alpha_{2,n}=\int_0^\pi\zeta_n^2(x)\,dx-\frac{1}{\lambda_n}\int_0^\pi p(x)\,\zeta_n^2(x)\,dx$$

where $\zeta(x,\lambda_n)=\zeta_n(x)$.
Now we set the spectral functions of problems (2.1)-(2.2) and (2.2), (2.6) as follows:

$$\rho_1(\lambda)=\begin{cases}\sum_{\lambda_{1,n}<\lambda}\frac{1}{\alpha_{1,n}}, & \lambda\geq 0\\ -\sum_{\lambda<\lambda_{1,n}}\frac{1}{\alpha_{1,n}}, & \lambda<0\end{cases}$$

$$\rho_2(\lambda)=\begin{cases}\sum_{\lambda_{2,n}<\lambda}\frac{1}{\alpha_{2,n}}, & \lambda\geq 0\\ -\sum_{\lambda<\lambda_{2,n}}\frac{1}{\alpha_{2,n}}, & \lambda<0\end{cases}$$

respectively.

3 Main results

In this section, our purpose is to give effective methods of restoring the diffusion equation from two spectra following the method for classical Sturm-Liouville operator and is to prove the stability theorem for diffusion equation.

Theorem 3.1. We have

$$\alpha_{1,n}=\frac{h_2-h_1}{2\lambda_{1,n}(\mu_{1,n}-\lambda_{1,n})}\prod_{k=-\infty}^{\infty}{}'\left(\frac{\lambda_{1,k}-\lambda_{1,n}}{\mu_{1,k}-\lambda_{1,n}}\right),\ (n=0,\pm1,\pm2,...),$$

where the symbols $\prod'$ denotes the number $k=n$ has been omitted from the infinite product.

Proof. Denote by $\varphi(x,\lambda)$ and $\psi(x,\lambda)$ the solutions of the equation (2.1) which satisfies the boundary conditions

$$\varphi(0,\lambda)=1\ ,\ \varphi'(0,\lambda)=h_1,$$
$$\psi(0,\lambda)=1\ ,\ \psi'(0,\lambda)=h_2.$$

The eigenvalues of boundary problems (2.1)-(2.2) and (2.1), (2.3) coincide with the zeros of the functions

$$\Phi_1(\lambda)=\varphi'(\pi,\lambda)+H\varphi(\pi,\lambda),$$
$$\Phi_2(\lambda)=\psi'(\pi,\lambda)+H\psi(\pi,\lambda),$$

respectively. Put

$$f(x,\lambda)=\psi(x,\lambda)+m(\lambda)\varphi(x,\lambda)$$

and require that $f(x,\lambda)$ should hold the condition

$$f'(\pi,\lambda)+Hf(\pi,\lambda)=0.$$

It follows that

$$m(\lambda)=-\frac{\psi'(\pi,\lambda)+H\psi(\pi,\lambda)}{\varphi'(\pi,\lambda)+H\varphi(\pi,\lambda)}=-\frac{\Phi_2(\lambda)}{\Phi_1(\lambda)}.$$

It can be seen from this formula that $m(\lambda)$ is a metamorphic function, its poles and zeros coincide with the eigenvalues of the problems (2.1)-(2.2) and (2.1), (2.3), respectively. On the other hand, it follows from Green's formula that

$$\left(\lambda_1^2-\lambda_2^2\right)\int_0^\pi f(x,\lambda_1)f(x,\lambda_2)\,dx=$$

$$\left(f'(x,\lambda_2)f(x,\lambda_1)-f'(x,\lambda_1)f(x,\lambda_2)\right)\Big|_0^\pi+\int_0^\pi f(x,\lambda_1)f(x,\lambda_2)(2\lambda_1-2\lambda_2)p(x)\,dx$$

$$=(m(\lambda_1)-m(\lambda_2))(h_1-h_2)+2(\lambda_1-\lambda_2)\int_0^\pi f(x,\lambda_1)f(x,\lambda_2)p(x)\,dx. \quad (3.1)$$

Putting $\lambda_1=\lambda,\ \lambda_2=\overline{\lambda}$ in (3.1) gives

$$\int_0^\pi |f(x,\lambda)|^2\,dx-\frac{1}{Re\lambda}\int_0^\pi p(x)|f(x,\lambda)|^2\,dx=\frac{(h_1-h_2)Imm(\lambda)}{2Im\lambda Re\lambda}.$$

It can be seen from this formula that if $h_1>h_2$ then the function $m(\lambda)$ maps the upper half-line onto itself (for $h_1<h_2$ this holds for the lower half-plane). Hence, the zeros and poles of the function $m(\lambda)$ alternates.

Applying Green's formula again, gives

$$\left(\lambda^2-\lambda_n^2\right)\int_0^\pi f(x,\lambda)\varphi(x,\lambda_n)\,dx=h_2-h_1+2(\lambda-\lambda_n)\int_0^\pi f(x,\lambda)\varphi(x,\lambda_n)p(x)\,dx,$$

or

$$\left(\lambda^2-\lambda_n^2\right)\int_0^\pi f(x,\lambda)\varphi(x,\lambda_n)\,dx-2(\lambda-\lambda_n)\int_0^\pi f(x,\lambda)\varphi(x,\lambda_n)p(x)\,dx=h_2-h_1.$$

On the other hand

$$\begin{aligned}&\left(\lambda^2-\lambda_n^2\right)\int_0^\pi \psi(x,\lambda)\varphi(x,\lambda_n)\,dx+\left(\lambda^2-\lambda_n^2\right)m(\lambda)\int_0^\pi \varphi(x,\lambda)\varphi(x,\lambda_n)\,dx\\&-2(\lambda-\lambda_n)\int_0^\pi \psi(x,\lambda)\varphi(x,\lambda_n)p(x)\,dx-2(\lambda-\lambda_n)m(\lambda)\int_0^\pi \varphi(x,\lambda)\varphi(x,\lambda_n)p(x)\,dx\\&=h_2-h_1.\end{aligned}$$

Assuming here that instead of all λ putting λ_n, $(\lambda\to\lambda_n)$

$$\begin{aligned}\alpha_n&=\int_0^\pi \varphi^2(x,\lambda_n)\,dx-\frac{1}{\lambda_n}\int_0^\pi p(x)\varphi^2(x,\lambda_n)\,dx\\&=\frac{h_2-h_1}{\lim\limits_{\lambda\to\lambda_n}2\lambda_n(\lambda-\lambda_n)m(\lambda)}.\end{aligned}\tag{3.2}$$

It is generally known that the functions of order 1/2 and are determined by their zeros up to constant factors thus:

$$\Phi_1(\lambda)=C_1\prod_{k=-\infty}^{\infty}\left(1-\frac{\lambda}{\lambda_k}\right),\quad \Phi_2(\lambda)=C_2\prod_{k=-\infty}^{\infty}\left(1-\frac{\lambda}{\mu_k}\right)$$

where C_1 and C_2 are constants. It follows that from these formulas and (3.2)

$$\alpha_n=\frac{h_2-h_1}{2\lambda_n(\mu_n-\lambda_n)}\frac{C_1}{C_2}\prod_{k=-\infty}^{\infty}\left(\frac{\mu_k}{\lambda_k}\right)\prod_{k=-\infty}^{\infty}{}'\left(\frac{\lambda_k-\lambda_n}{\mu_k-\lambda_n}\right).$$

Now it has to be shown that

$$\frac{C_1}{C_2}\prod_{k=-\infty}^{\infty}\left(\frac{\mu_k}{\lambda_k}\right)=1.\tag{3.3}$$

Note that it follows from the asymptotics formulas for a solution of the diffusion equation that

$$\begin{aligned}&\Phi_1(\lambda)=\varphi'(\pi,\lambda)+H\varphi(\pi,\lambda)\\&=\left(\alpha'_{h_1}(\pi)-\lambda\right)\sin\left(\lambda\pi-\alpha_{h_1}(\pi)\right)+A_{h_1}(\pi,\pi)\cos\lambda\pi\\&+B_{h_1}(\pi,\pi)\sin\lambda\pi+H\cos\left(\lambda\pi-\alpha_{h_1}(\pi)\right)\\&+\int_0^\pi\left(HA_{h_1}(\pi,t)+\left.\frac{\partial A_{h_1}(x,t)}{\partial x}\right|_{x=\pi}\right)\cos\lambda t\,dt\\&+\int_0^\pi\left(HB_{h_1}(\pi,t)+\left.\frac{\partial B_{h_1}(x,t)}{\partial x}\right|_{x=\pi}\right)\sin\lambda t\,dt\end{aligned}$$

and

$$\begin{aligned}&\Phi_2(\lambda)=\psi'(\pi,\lambda)+H\psi(\pi,\lambda)\\&=\left(\alpha'_{h_2}(\pi)-\lambda\right)\sin\left(\lambda\pi-\alpha_{h_2}(\pi)\right)+A_{h_2}(\pi,\pi)\cos\lambda\pi\\&+B_{h_2}(\pi,\pi)\sin\lambda\pi+H\cos\left(\lambda\pi-\alpha_{h_2}(\pi)\right)\\&+\int_0^\pi\left(HA_{h_2}(\pi,t)+\left.\frac{\partial A_{h_2}(x,t)}{\partial x}\right|_{x=\pi}\right)\cos\lambda t\,dt\\&+\int_0^\pi\left(HB_{h_2}(\pi,t)+\left.\frac{\partial B_{h_2}(x,t)}{\partial x}\right|_{x=\pi}\right)\sin\lambda t\,dt.\end{aligned}$$

The last two equalities yield that

$$\lim_{\lambda\to-\infty}\frac{\Phi_1(\lambda)}{\Phi_2(\lambda)}=1$$

that is

$$\lim_{\lambda\to-\infty}\frac{C_1}{C_2}\prod_{k=-\infty}^{\infty}\left(1-\frac{\lambda}{\lambda_k}\right)\left(1-\frac{\lambda}{\mu_k}\right)^{-1}$$

$$=\frac{C_1}{C_2}\prod_{k=-\infty}^{\infty}\frac{\mu_k}{\lambda_k}\lim_{\lambda\to-\infty}\prod_{k=-\infty}^{\infty}\frac{\lambda_k-\lambda}{\mu_k-\lambda}=1.$$

We now prove that

$$\lim_{\lambda\to-\infty}\prod_{k=-\infty}^{\infty}\frac{\lambda_k-\lambda}{\mu_k-\lambda}=\lim_{\lambda\to-\infty}\prod_{k=-\infty}^{\infty}\left(1+\frac{\lambda_k-\mu_k}{\mu_k-\lambda}\right)=1. \tag{3.4}$$

From the asymptotic formulas for the eigenvalues of the considered problem, it is known that $\lambda_k = n + c_0 + O(1)$ and $\mu_k = n + c_0 + O(1)$. Hence $\lambda_k - \mu_k = O(1)$ and the series

$$\sum_{k=-\infty}^{\infty}\frac{\lambda_k-\mu_k}{\mu_k-\lambda}=\sum_{k=1}^{\infty}\left(\frac{\lambda_k-\mu_k}{\mu_k-\lambda}+\frac{\lambda_{-k}-\mu_{-k}}{\mu_{-k}-\lambda}\right)+\frac{\lambda_0-\mu_0}{\mu_0-\lambda}.$$

converges uniformly in a neighbourhood of the point $\lambda=-\infty$. Therefore the limit in each term of the infinite product (3.4) can be approached, that is,

$$\lim_{\lambda\to-\infty}\prod_{k=-\infty}^{\infty}\left(1+\frac{\lambda_k-\mu_k}{\mu_k-\lambda}\right)=1$$

and the Eq. (3.3) is obtained from the last equality. Thereby the proof is completed. Q.E.D.

Then the expressions of norming constants in terms of two spectra of problems (2.1)-(2.2) and (2.1), (2.3), respectively, are shown by

$$\alpha_{1,n}=\frac{h_2-h_1}{2\lambda_{1,n}\left(\mu_{1,n}-\lambda_{1,n}\right)}\prod_{k=-\infty}^{\infty}{}'\left(\frac{\lambda_{1,k}-\lambda_{1,n}}{\mu_{1,k}-\lambda_{1,n}}\right),\quad (n=0,\pm1,\pm2,...)\,,$$

$$\alpha_{2,n}=\frac{h_2-h_1}{2\lambda_{2,n}\left(\mu_{2,n}-\lambda_{2,n}\right)}\prod_{k=-\infty}^{\infty}{}'\left(\frac{\lambda_{2,k}-\lambda_{2,n}}{\mu_{2,k}-\lambda_{2,n}}\right),\quad (n=0,\pm1,\pm2,...)\,.$$

Now we give main theorem in this study.

Theorem 3.2. Let the eigenvalues $\{\lambda_{j,k}\}$ and $\{\mu_{j,k}\}$ $(j=1,2)$ coincide the numbers of $2N+2$ such that $\lambda_{1,k}=\lambda_{2,k}$ and $\mu_{1,k}=\mu_{2,k}$ for $k=1,2,...,N+1$ and $k=-1,...,-N-1$ then

$$\underset{\substack{-\infty<\lambda<\frac{N}{2}\\ -\frac{N}{2}<\lambda<\infty}}{Var}\{\rho_1(\lambda)-\rho_2(\lambda)\}\le\begin{cases} Ce^{C}\rho_1\left(\frac{N}{2}\right) &, \lambda>0\\ \tilde{C}e^{\tilde{C}}\rho_2\left(-\frac{N}{2}\right) &, \lambda<0\end{cases}$$

respectively, for $k>N+1$, $n<\frac{N}{2}$, $N>2A$ and $N>2\,\tilde{A}$ in case of $\lambda>0$ and for $k<-N-1$, $n>-\frac{N}{2}$, $N>2A$ and $N>2\,\tilde{A}$ in case of $\lambda<0$, where

$$A=\left|\frac{1}{2\pi k}\int_0^\pi\left(q_2\left(t\right)-q_1\left(t\right)\right)dt+\frac{c_{2,k}-c_{1,k}}{k}\right|,$$

$$\tilde{A}=\left|\frac{1}{2\pi k}\int_0^\pi\left(q_1\left(t\right)-q_2\left(t\right)\right)dt+\frac{\tilde{c}_{1,k}-\tilde{c}_{2,k}}{k}\right|,$$

$C=\frac{\left(1+\frac{3}{2N}\right)\left(A+\tilde{A}\right)}{N\left(\frac{1}{2}+\frac{7}{4N}+\frac{3}{2N^2}\right)}$ and $\tilde{C}=\frac{\left(1+\frac{5}{2N}\right)\left(A+\tilde{A}\right)}{N\left(\frac{1}{2}+\frac{17}{4N}+\frac{15}{2N^2}\right)}$.

Proof. From the definition of the variation, it follows that

$$\underset{\substack{-\infty<\lambda<\lambda_0\\ \lambda_{\tilde{0}}<\lambda<\infty}}{Var}\left\{\rho_1\left(\lambda\right)-\rho_2\left(\lambda\right)\right\}\le\begin{cases}\max\limits_{\lambda_n<\lambda_0}\left|1-\frac{\alpha_{1,n}}{\alpha_{2,n}}\right|\sum_{\lambda_n<\lambda_0}\frac{1}{\alpha_{1,n}}, & \lambda\ge 0\\ \max\limits_{\lambda_{\tilde{0}}<\lambda_n}\left|1-\frac{\alpha_{2,n}}{\alpha_{1,n}}\right|\sum_{\lambda_{\tilde{0}}<\lambda_n}\frac{1}{\alpha_{2,n}}, & \lambda<0\end{cases}$$

$$\le\begin{cases}\max\limits_{\lambda_n<\lambda_0}\left|1-\frac{\alpha_{1,n}}{\alpha_{2,n}}\right|\rho_1\left(\lambda_{N+2}\right), & \lambda\ge 0\\ \max\limits_{\lambda_{\tilde{0}}<\lambda_n}\left|1-\frac{\alpha_{2,n}}{\alpha_{1,n}}\right|\rho_2\left(\lambda_{-N-2}\right), & \lambda<0\end{cases}\tag{3.5}$$

for $\lambda_0<\lambda_{N+2}$ and $\lambda_{-N-2}<\lambda_{\tilde{0}}$, respectively. The assessment of the variation of the spectral functions is reduced to evaluation of the absolute value on the right part of (3.5).
So firstly, we must examine the inequality

$$\max_{n<\frac{N}{2}}\left|1-\frac{\alpha_{1,n}}{\alpha_{2,n}}\right|\le\max_{n<\frac{N}{2}}\left|1-\prod_{k=N+2}^{\infty}\frac{\left(\lambda_{1,k}-\lambda_{1,n}\right)\left(\mu_{2,k}-\lambda_{2,n}\right)}{\left(\lambda_{2,k}-\lambda_{2,n}\right)\left(\mu_{1,k}-\lambda_{1,n}\right)}\right|.\tag{3.6}$$

In this case, the eigenvalues coincide only for $k=1,2,...,N+1$, that is $\lambda_{1,k}=\lambda_{2,k}$ and $\mu_{1,k}=\mu_{2,k}$ for $k=1,2,...,N+1$. Considering the infinite product

$$\Psi_1\left(\lambda_n\right)=\prod_{k=N+2}^{\infty}\frac{\left(\lambda_{1,k}-\lambda_{1,n}\right)\left(\mu_{2,k}-\lambda_{2,n}\right)}{\left(\lambda_{2,k}-\lambda_{2,n}\right)\left(\mu_{1,k}-\lambda_{1,n}\right)}\tag{3.7}$$

and taking the logarithm of each side of (3.7), it can be easily obtained that

$$\left|\ln\Psi_1\left(\lambda_n\right)\right|=\left|\sum_{k=N+2}^{\infty}\ln\left(\frac{\lambda_{1,k}-\lambda_{1,n}}{\lambda_{2,k}-\lambda_{2,n}}\right)+\sum_{k=N+2}^{\infty}\ln\left(\frac{\mu_{2,k}-\lambda_{2,n}}{\mu_{1,k}-\lambda_{1,n}}\right)\right|,$$

$$\le\sum_{k=N+2}^{\infty}\left|\ln\left(1-\frac{\lambda_{2,k}-\lambda_{1,k}}{\lambda_{2,k}-\lambda_{1,n}}\right)\right|+\sum_{k=N+2}^{\infty}\left|\ln\left(1-\frac{\mu_{1,k}-\mu_{2,k}}{\mu_{1,k}-\lambda_{2,n}}\right)\right|.\tag{3.8}$$

For $k>N+1$, $n<\frac{N}{2}$, it follows that $\left|\frac{\lambda_{2,k}-\lambda_{1,k}}{\lambda_{2,k}-\lambda_{1,n}}\right|<1$ and $\left|\frac{\mu_{1,k}-\mu_{2,k}}{\mu_{1,k}-\lambda_{2,n}}\right|<1$.

From the last inequalities Eq. (3.8) can also be written as

$$|\ln \Psi_1(\lambda_n)| < \sum_{k=N+2}^{\infty} \frac{\left|\frac{\lambda_{2,k}-\lambda_{1,k}}{\lambda_{2,k}-\lambda_{1,n}}\right|}{1-\left|\frac{\lambda_{2,k}-\lambda_{1,k}}{\lambda_{2,k}-\lambda_{1,n}}\right|} + \sum_{k=N+2}^{\infty} \frac{\left|\frac{\mu_{1,k}-\mu_{2,k}}{\mu_{1,k}-\lambda_{2,n}}\right|}{1-\left|\frac{\mu_{1,k}-\mu_{2,k}}{\mu_{1,k}-\lambda_{2,n}}\right|}. \tag{3.9}$$

Considering asymptotic formulas of the eigenvalues (2.4), (2.5), (2.7), (2.8) and after some straightforward computations we can obtain that

$$\begin{aligned} \left|\frac{\lambda_{2,k}-\lambda_{1,k}}{\lambda_{2,k}-\lambda_{1,n}}\right| &= \left|\frac{\frac{1}{2\pi k}\int_0^\pi (q_2(t)-q_1(t))\,dt + \frac{c_{2,k}-c_{1,k}}{k}}{\lambda_{2,k}\left(1-\frac{\lambda_{1,n}}{\lambda_{2,k}}\right)}\right| \\ &< \frac{A}{\left(k-\frac{1}{2}\right)\left(1-\frac{\lambda_{1,\left[\!\left[\frac{N}{2}\right]\!\right]}}{\lambda_{2,N+2}}\right)} < \frac{A\left(1+\frac{3}{2N}\right)}{N\left(\frac{1}{2}+\frac{7}{4N}+\frac{3}{2N^2}\right)} \end{aligned} \tag{3.10}$$

and

$$\begin{aligned} \left|\frac{\mu_{1,k}-\mu_{2,k}}{\mu_{1,k}-\lambda_{2,n}}\right| &< \frac{\left|\frac{1}{2\pi k}\int_0^\pi (q_1(t)-q_2(t))\,dt + \frac{\tilde{c}_{1,k}-\tilde{c}_{2,k}}{k}\right|}{\left|\left(k-\frac{1}{2}\right)\left(1-\frac{\lambda_{2,\left[\!\left[\frac{N}{2}\right]\!\right]}}{\mu_{1,N+2}}\right)\right|} \\ &< \frac{\tilde{A}\left(1+\frac{3}{2N}\right)}{N\left(\frac{1}{2}+\frac{7}{4N}+\frac{3}{2N^2}\right)}. \end{aligned} \tag{3.11}$$

Substituting (3.10) and (3.11) into (3.9), it follows that for $N > 2A$ and $N > 2\tilde{A}$

$$|\ln \Psi_1(\lambda_n)| < \frac{\left(1+\frac{3}{2N}\right)\left(A+\tilde{A}\right)}{N\left(\frac{1}{2}+\frac{7}{4N}+\frac{3}{2N^2}\right)}. \tag{3.12}$$

Considering (3.6) with (3.12) and using serial expansion of exponential function then the proof of Theorem 3.2 is completed for positive eigenvalues.
On the other hand, we consider that

$$\max_{n>-\frac{N}{2}}\left|1-\frac{\alpha_{2,n}}{\alpha_{1,n}}\right| = \max_{n>-\frac{N}{2}}\left|1-\prod_{k=-\infty}^{-N-2}\frac{(\lambda_{2,k}-\lambda_{2,n})(\mu_{1,k}-\lambda_{1,n})}{(\lambda_{1,k}-\lambda_{1,n})(\mu_{2,k}-\lambda_{2,n})}\right|. \tag{3.13}$$

In this case, the eigenvalues coincide only for $k=-1,...,-N-1$, that is $\lambda_{1,k}=\lambda_{2,k}$ and $\mu_{1,k}=\mu_{2,k}$ for $k=-1,...,-N-1$. Considering the infinite product

$$\Psi_2(\lambda_n) = \prod_{k=-\infty}^{-N-2}\frac{(\lambda_{2,k}-\lambda_{2,n})(\mu_{1,k}-\lambda_{1,n})}{(\lambda_{1,k}-\lambda_{1,n})(\mu_{2,k}-\lambda_{2,n})},$$

by making similar calculations as above it follows that

$$|\ln \Psi_2(\lambda_n)| = \left|\sum_{k=-\infty}^{-N-2}\ln\left(\frac{\lambda_{2,k}-\lambda_{2,n}}{\lambda_{1,k}-\lambda_{1,n}}\right) + \sum_{k=-\infty}^{-N-2}\ln\left(\frac{\mu_{1,k}-\lambda_{1,n}}{\mu_{2,k}-\lambda_{2,n}}\right)\right|$$

$$\leq \sum_{k=-\infty}^{-N-2} \left| \ln\left(1 - \frac{\lambda_{1,k} - \lambda_{2,k}}{\lambda_{1,k} - \lambda_{2,n}}\right) \right| + \sum_{k=-\infty}^{-N-2} \left| \ln\left(1 - \frac{\mu_{2,k} - \mu_{1,k}}{\mu_{2,k} - \lambda_{1,n}}\right) \right|. \tag{3.14}$$

for $k < -N - 1$ and $n > -\frac{N}{2}$, it can be easily seen $\left|\frac{\lambda_{2,k}-\lambda_{1,k}}{\lambda_{2,k}-\lambda_{1,n}}\right| < 1$ and $\left|\frac{\mu_{1,k}-\mu_{2,k}}{\mu_{1,k}-\lambda_{2,n}}\right| < 1$. From the last inequalities rewrite Eq. (3.14) as

$$|\ln \Psi_2(\lambda_n)| < \sum_{k=-\infty}^{-N-2} \frac{\left|\frac{\lambda_{1,k}-\lambda_{2,k}}{\lambda_{1,k}-\lambda_{2,n}}\right|}{1 - \left|\frac{\lambda_{1,k}-\lambda_{2,k}}{\lambda_{1,k}-\lambda_{2,n}}\right|} + \sum_{k=-\infty}^{-N-2} \frac{\left|\frac{\mu_{2,k}-\mu_{1,k}}{\mu_{2,k}-\lambda_{1,n}}\right|}{1 - \left|\frac{\mu_{2,k}-\mu_{1,k}}{\mu_{2,k}-\lambda_{1,n}}\right|}. \tag{3.15}$$

Here using the classical asymptotics of the eigenvalues, it follows that

$$\begin{aligned} \left|\frac{\lambda_{1,k} - \lambda_{2,k}}{\lambda_{1,k} - \lambda_{2,n}}\right| &< \left|\frac{\frac{1}{2\pi k}\int_0^\pi (q_2(t) - q_1(t))\, dt + \frac{c_{2,k}-c_{1,k}}{k}}{\left(-k - \frac{1}{2}\right)\left(1 - \frac{\lambda_{2,\left[\!\left[-\frac{N}{2}\right]\!\right]}}{\lambda_{1,-N-2}}\right)}\right| \\ &< \frac{A\left(1 + \frac{5}{2N}\right)}{N\left(\frac{1}{2} + \frac{17}{4N} + \frac{15}{2N^2}\right)} \end{aligned} \tag{3.16}$$

and

$$\begin{aligned} \left|\frac{\mu_{2,k} - \mu_{1,k}}{\mu_{2,k} - \lambda_{1,n}}\right| &< \frac{\left|\frac{1}{2\pi k}\int_0^\pi (q_1(t) - q_2(t))\, dt + \frac{\tilde{c}_{1,k}-\tilde{c}_{2,k}}{k}\right|}{\left|\left(-k - \frac{1}{2}\right)\left(1 - \frac{\lambda_{1,\left[\!\left[-\frac{N}{2}\right]\!\right]}}{\mu_{2,-N-2}}\right)\right|} \\ &< \frac{\tilde{A}\left(1 + \frac{5}{2N}\right)}{N\left(\frac{1}{2} + \frac{17}{4N} + \frac{15}{2N^2}\right)}. \end{aligned} \tag{3.17}$$

Substituting (3.16) and (3.17) into (3.15), we have for $N > 2A$ and $N > 2\tilde{A}$

$$\begin{aligned} |\ln \Psi_2(\lambda_n)| &< \frac{\frac{A\left(1+\frac{5}{2N}\right)}{N\left(\frac{1}{2}+\frac{17}{4N}+\frac{15}{2N^2}\right)}}{1 - \frac{A\left(1+\frac{5}{2N}\right)}{N\left(\frac{1}{2}+\frac{17}{4N}+\frac{15}{2N^2}\right)}} + \frac{\frac{\tilde{A}\left(1+\frac{5}{2N}\right)}{N\left(\frac{1}{2}+\frac{17}{4N}+\frac{15}{2N^2}\right)}}{1 - \frac{\tilde{A}\left(1+\frac{5}{2N}\right)}{N\left(\frac{1}{2}+\frac{17}{4N}+\frac{15}{2N^2}\right)}} \\ &< \frac{\left(A + \tilde{A}\right)\left(1 + \frac{5}{2N}\right)}{N\left(\frac{1}{2} + \frac{17}{4N} + \frac{15}{2N^2}\right)}. \end{aligned} \tag{3.18}$$

Then considering (3.13) with (3.18) and using serial expansion of exponential function we can obtain proof of Theorem 3.2 for negative eigenvalues. Therefore the proof is completed. Q.E.D.

4 Conclusion

In spectral theory, the stability theorem explains the stability of solutions or spectral functions of differential equations under small perturbations of initial conditions. For instance, if small perturbations of initial data of considered differential equation lead to small variations then differential equation is stable. In this study, applying Ryabushko's method, we obtain the proximity of spectral functions of problems (2.1)-(2.2) and (2.2), (2.6) whose eigenvalues $\{\lambda_{j,k}\}$ and $\{\mu_{j,k}\}(j = 1, 2)$ coincide numbers of $2N + 2$. Also, the procedure of reconstruction of the equation (2.1) from its two spectra is indicated.

References

[1] M. G. Gasymov and G. Sh. Guseinov, *Determination of a diffusion operator from spectral data*, Akademiya Nauk Azerbaĭdzhansko ĭ SSR. Doklady, **37(2)** (1981) 19–23.

[2] B. M. Levitan and I. S. Sargsjan, *Introduction to spectral theory*, American Mathematical Society, Providence, RI, USA, (1975).

[3] B. M. Levitan and M. G. Gasymov, *Determination of a differential equation by two spectra*, Akademiya Nauk SSSR i Moskovskoe Matematicheskoe Obshchestvo, **19** (1964) 3-63.

[4] B. M. Levitan, *On the determination of the Sturm–Liouville operator from one and two spectra*, Mathematics of the USSR Izvestija., **12** (1978) 179–193.

[5] V. A. Yurko, *An inverse problem for pencils of differential operators*, Matematicheskiĭ Sbornik, **191(10)** (2000), 137–160.

[6] M. Jaulent and C. Jean *The inverse s-wave scattering problem for a class of potentials depending on energy*, Communications in Mathematical Physics, **28** (1972) 177–220.

[7] G. Sh. Guseinov, *On spectral analysis of a quadratic pencil of Sturm-Liouville operators*, Soviet Mathematics Doklady, **32(3)** (1985) 859–862.

[8] I. M. Guseinov and I. M. Nabiev, *The inverse spectral problem for pencils of differential operators*, Sbornik Mathematics, **198(11)** (2007) 1579–1598.

[9] G. Sh. Guseinov, *On construction of a quadratic Sturm-Liouville operator pencil from spectral data*, Proceeding of IMM of NAS of Azerbaijan, **40** (2014) 203–214.

[10] R. Hryniv and N. Pronska, *Inverse spectral problems for energy-dependent Sturm-Liouville equations*, Inverse Problems, **28(8)** (2012) 085008.

[11] R. Kh. Amirov and A. A. Nabiev, *Inverse problems for the quadratic pencil of the Sturm-Liouville equations with impulse*, Abstract and Applied Analysis, Article ID 361989, (2013).

[12] H. Koyunbakan and E. S. Panakhov, *Half-inverse problem for diffusion operators on the finite interval*, Journal of Mathematical Analysis and Applications, **326(2)** (2007) 685–692.

[13] T. Gulsen and E.S. Panakhov, *On the isospectrality of the scalar energy-dependent Schrödinger problems*, Turkish Journal of Mathematics, **42** (2018) 139–154.

[14] M. Sat and E. S. Panakhov, *Spectral problem for diffusion operator*, Applicable Analysis, **93(6)** (2014) 1178–1186.

[15] E. S. Panakhov and R. Yilmazer, *On inverse problem for singular Sturm-Liouville operator from two spectra*, Ukrainian Mathematical Journal, **58(1)** (2006) 147–154.

[16] E. Bas and E. S. Panakhov, *A new approximation for singular inverse Sturm-Liouville Problem*, Thai Journal of Mathematics, **10(3)** (2012) 1024–1030.

[17] V. A. Marchenko, *Spectral theory of Sturm-Liouville operators*, Naukova Dumka, Kiev, (1972) 187–207.

[18] V. A. Marchenko and K. V. Maslov, *Stability of the problem of recovering the Sturm-Liouville operator from the spectral function*, Mathematics of the USSR Sbornik, **81(123)** (1970) 475–502.

[19] J. R. McLaughlin, *Stability theorems for two inverse spectral problems*, Inverse Problems, **4** (1988) 529–540.

[20] A. M. Savchuk and A. A. Shkalikov, *Inverse problems for Sturm–Liouville operators with potentials in Sobolev spaces: Uniform stability*, Functional Analysis and Its Applications, **44** (2010) 270–285.

[21] A. Ercan and E. S. Panakhov, *Stability of the reconstruction discontinuous Sturm-Liouville problem,* Communications Faculty of Sciences University of Ankara Series A1 Mathematics and Statistics, **68(1)** (2019) 484–499.

[22] E. S. Panakhov and A. Ercan, *Stability problem of singular Sturm-Liouville equation,* TWMS Journal of Pure and Applied Mathematics, **8(2)** (2017) 148–159.

[23] T. I. Ryabushko, *Stability of the reconstruction of a Sturm–Liouville operator from two spectra, II*, Teor. Funkts. Anal. Prilozhen, **18** (1973) 176–185 (in Russian).

[24] T. I. Ryabushko, *Stability of the reconstruction of a Sturm–Liouville boundary value problem on a semiaxis from two spectra*, Teor. Funktsiĭ Funktsional. Anal. Prilozhen **35** (1981) 96–100.

[25] A. Ercan, *The stability problems in spectral theory of differential operators,* Ph.D. Thesis, Firat University, 2018.

DOI 10.2478/9788395793882-011

New ways for building BCK-algebras of higher order

Radu Vasile*

Ovidius University, Bd. Mamaia 124, Constanţa, 900527, România
E-mail: rvasile@gmail.com

Abstract

In this paper we present new extensions for BCK-algebras. Also, we give a representation of the BCK-algebras class structure, by introducing a new concept, BCK-trees.

2010 Mathematics Subject Classification. **06F35** 18B35
Keywords. BCK-algebras, BCK-algebra extension, BCK-tree.

1 Introduction

BCK-algebras were first introduced in 1966 by Y. Imai and K. Iseki, in [4], as a generalization of the difference operation of two sets and generalization of propositional calculi. For example, the relation

$$A - (A - B) \subset B, A, B \text{ sets},$$

in propositional calculi can be written

$$p \rightarrow ((p \rightarrow q) \rightarrow q).$$

Recently, there were revealed new interesting applications of these new algebras, such as those found in [5] and [3]. In the first of the two articles, a "block code" is constructed from a BCK-algebra. At the end of the article, the authors ask if the reverse procedure is possible. In the second quoted article, a partial answer is given to this question, showing that, under certain conditions, the answer is affirmative, it is possible to construct a BCK-algebra starting from a "block code". Surprisingly, however, the BCK-algebra obtained from the code is not isomorphic with the BCK-algebra from which the same code was obtained, which means that code can have 2 associated BCK-algebras. In the article [3] it is shown that these algebras are in an equivalence relation called "code similar". Moreover, it is shown that the two algebras do not even have the same properties, one being commutative and non-implicative, the other being non-commutative and non-implicative. On a closer look, it can be noticed that in the development of the codes from BCK-algebras, or, of the BCK-algebras from codes, what is actually used is the table of operations that the algebra generates. The definition of BCK-algebra gives us the uniqueness of the codewords, obtained by applying some transformations on the operations table of that algebra. Since what counts in getting the codewords is the operations table, we can have isomorphic algebras, with different codes attached. Hence, it is important to study all BCK-algebras, not just the equivalence class representatives from the isomorphism relation. For this, it is important to see how we can construct BCK-algebras directly, by computations, or by the help of the computer and to isolate various sets of BCK-algebras, based on their properties. In the process of constructing such an algebra, we want to obtain an algebra with certain properties, or, when we talk about the construction of new BCK-algebras, by extension from some existing ones, we want not to lose any properties.

*Special big thanks to conf. univ. dr. Cristina Flaut, Ovidius University, Constanţa, România

Tbilisi Mathematical Journal Special Issue (IECMSA-2019), pp. 111–124.
Tbilisi Centre for Mathematical Sciences.

Received by the editors: 15 March 2020.
Accepted for publication: 24 April 2020.

2 Preliminaries

2.1 BCK-algebras

Definition 2.1 ([6, 7]). A set $(X, *, 0)$ of type $(2, 0)$ is called a *BCI*-algebra if the following conditions are satisfied:

1. $((x * y) * (x * z)) * (z * y) = 0$, $\forall x, y, z \in X$;
2. $(x * (x * y)) * y = 0$, $x, y \in X$;
3. $x * x = 0$, $\forall x \in X$;
4. If $x * y = 0$ and $y * x = 0$, then $x = y, \forall x, y \in X$.

Definition 2.2 ([6, 7]). If a BCI-algebra X satisfies the following relations:

5. $0 * x = 0$, $\forall x \in X$,

then X is called a *BCK*-algebra.

In a BCI-algebra, the following relations also apply [7, pp 6] and [6]:

6. $(x * y) * z = (x * z) * y$;
7. $(x * y) * x = 0$;
8. $((x * z) * (y * z)) * (x * y) = 0$;
9. $x * 0 = x$;

Some authors give an alternative BCI-algebra definition, replacing axioms from the original definition with one of those listed above, proving that it is also a well-defined axiomatic system for BCI-algebra [see [2] prop. 3.1]:

Proposition 2.3. A BCI-algebra is characterized by the following axioms:
1, 4, 9,
and a BCK-algebra through the axioms:
1, 4, 9, 5.

Definition 2.4 ([3]). If $(X, *, 0)$ and $(Y, \circ, 0)$ are two BCK-algebras, an application $f : X \to Y$ with the property $f(x * y) = f(x) \circ f(y)$, for any $x, y \in X$, is called a BCK-algebras morphism. If f is a bijective application, then f is a *isomorphism* of BCK-algebras.

Remark 2.5. It can be observed that we can define a partial order relation $<$ on a BCK-algebra by the formula $x < y$ if and only if $x * y = 0$.

2.2 Properties of BCI / BCK-algebras

Definition 2.6 ([6]). In a BCI / BCK-algebra X if there is a generic element denoted with 1 that satisfies the relation $x < 1, \forall x \in X$, element 1 is called the unit of X algebra. A BCI / BCK-algebra with unit is called a bounded algebra.

Definition 2.7. [6] In a bounded BCK-algebra, an element with the property $1 * (1 * x) = x$ is called an involution.

Definition 2.8. A BCI / BCK-algebra X is called *commutative* if $x * (x * y) = y * (y * x)$, any $x, y \in X$.

Definition 2.9. A BCI / BCK-algebra x is called *implicative* if $x * (y * x) = x$, any $x, y \in X$.

Definition 2.10. A BCI / BCK-algebra $X, *, 0$ is called positive-implicative, if it satisfies the relation: $(x * z) * (y * z) = (x * y) * z$, for all x , $y, z \in X$.

Remark 2.11 ([2]). A BCI-Commutative Algebra is a commutative BCK-Algebra.

Remark 2.12. A BCI-algebra is implicative if it is commutative and positive-implicative.

In [6], the Iseki extension of a BCK-algebra is presented. This extension is a way to obtain new higher-order BCK-algebras from lower-order ones and is defined as follows:

Definition 2.13. Let B be a BCK-algebra. This can be extended to a Bounded BCK-algebra by adding a 1 element that complies with the following rules:

1. $1 * 1 = x * 1 = 0, \forall x \in B$.
2. $1 * x = 1, \forall x \in B$.

Remark 2.14. If a BCK-algebra B is a positive-implicative BCK-algebra, then the Iseki extension is a positive-implicative BCK-algebra.

Remark 2.15. If a BCK-algebra B is commutative, the Iseki Extension is not commutative. Considering that it does not preserve commutativity, the extension of Iseki does not preserve the property of being implicative.

3 Main results

3.1 New types of extensions for BCK-algebras

In the previous section, it is seen that BCK-algebras can be constructed by extension from the order n to the order $n+1$ through the Iseki extension. This is defined by introducing a new element and extending the operation from the source algebra to the resulting algebra. In the following, we build new extensions for BCK-algebras.

3.1.1 The Pseudo-Iseki Extension

Proposition 3.1. Let $(X, *, 0)$ be a BCK-algebra of order n and an element d that does not belong to X with the following properties:

1. $d * d = 0$;

2. $0 * d = 0$;

3. $d * x = d, \forall x \in X$;

4. $x * d = x, \forall x \in X$;

Then the algebra $X[d]$ is a BCK-algebra.

Proof. To be a BCK-algebra, the newly constructed algebra must satisfy the system of axioms. From the way we chose the element d, we have $d * d = 0$, $0 * d = 0$ and $d * x = d$, $x * d = x$, whatever $x \in X$. Thus, the axioms 3, 4, 5 in the axiomatic system of BCK-algebras are fulfilled. For axiom 2 we have $(d * (d * y)) * y = 0, \forall y \in X$ because $d * y = d$ and $d * d = 0$. On the other hand $(x * (x * d)) * d = 0, \forall x \in X$ because $x * d = x$ and $x * x = 0$.
We still have axiom 1 of the axiomatic system. $((d * y) * (d * z)) * (z * y) = 0$ because the first two brackets give d, and $d * d = 0$.
$((x * d) * (x * z)) * (z * d) = 0$ because $x * d = x$ and $z * d = z$, and the expression is reduced to: $(x * (x * z)) * z$ which is 0 from the definition of BCK-algebra X.
$((d * y) * (d * d)) * (d * y) = 0$ because the middle bracket is canceled, and the extreme brackets are similar. $((x * y) * (x * d)) * (d * y) = 0$ because: $x * d = x$, $d * y = d$, so the expression is reduced to $((x * y) * x) * d$. By the definition of the element d, the expression is reduced to $(x * y) * x$, this being 0 according to relation 7 in Definition 2.2. Q.E.D.

Definition 3.2. The extension from the previous proposition will be called pseudo-Iseki extension.

Proposition 3.3. If the source algebra is commutative, by pseudo-Iseki extension a commutative algebra can be obtained. If the source algebra is a positive-implicative one, the algebra obtained by pseudo-Iseki extension is a positive-implicative one.

Proof. Let $(X, *, 0)$ be a commutative and positive-implicative BCK-algebra. For commutativity, we have $x * (x * d) = d * (d * x)$. $(d * x) = d$, so $d * (d * x)) = 0$. $(x * d) = x$, so $x * (x * d)) = 0$. On the other hand, we have $d * (d * y) = y * (y * d)$. $(d * y) = d$, so $d * (d * y) = 0$. $(y * d) = y$, so, $y * (y * d) = 0$. The case where all the elements are d is trivial, since $d * d = 0$, and $d * 0 = d$, the final relation being $d = d$. For positive-implicative we have the following expressions: $(d * z) * (y * z) = (d * y) * z$. $(d * z) = d$, whence the expression becomes $d * (y * z) = (d * y) * z$. $(d * y) = d$, and $d * z = d$. so, the expression is $d * (y * z) = d$, which is true from the definition of the element d.
$(x * d) * (y * d) = (x * y) * d$. The parentheses from the left hand expression are x respectively y, and the right hand expression is $x * y$, so the entire expression is true. Last expression to be checked: $(x * z) * (d * z) = (x * d) * z$. $(d * z) = d$, $(x * d) = x$ and the expression becomes $(x * z) * d = x * z$ which is true from the definition of the element d . Q.E.D.

Remark 3.4. Regarding the order relation defined on the BCK-algebra, the pseudo-Iseki extension introduces a new element that does not compare with any of the elements in the source algebra.

3.1.2 The involution extension

In the following, we propose an interesting new extension. We propose it for research because it cannot be applied to all BCK-algebras, but only to certain algebras, thus defining a interesting type of BCK-algebras. Based on this extension, we can derive a construction process, applied to the operation tables of some BCK-algebras, to obtain higher-order BCK-algebras. Interesting to this process is that it does not affect the commutativity property and in some cases, it does not affect the positive-implicative property either.

Proposition 3.5. Let $(X, *, 0)$ be a BCK-algebra of order n and $d \notin X$. Respecting the following relations, we can obtain a BCK-algebra of order $n+ 1$:

1. $x * d = 0$ any $x \in X$;

2. $d * d = 0$;

3. $d * x = y$ if and only if $d * y = x$ any $x, y \in X$;

4. $((d * y) * (d * z)) * (z * y) = 0$ any $x, y, z \in X$.

Proof. From the choice of the element d, the axioms 3, 4, 5 of the axiomatic system of BCK-algebras are respected. For the axiom 1, regardless of the expression from 4, the only problematic relation remains: $((x*y)*(x*d))*(d*y) = 0$. But $(x*d) = 0$, so the expression reduces at $(x*y)*(d*y) = 0$. The element $(d * y) \in X)$ and $x = (d * z)$, $z \in X$, hence we can write: $((d * z) * y) * (d * y) = 0$. Applying a transformation we obtain: $((d * z) * (d * y)) * y = 0$. This expression should be true for any $y, z \in X$, therefore we can exchange y with z and the relation becomes: $((d*y)*(d*z))*z = 0$. In the relation 4 we have $((d * y) * (d * z)) * (z * y) = 0$. This means that $(d * y) * (d * z) < (z * y)$, but $z * y < z$, whence $(d * y) * (d * z) < z$ this mean $((d * y) * (d * z)) * z = 0$. Axiom 2 remains to be verified. For the 2 axiom we have the cases:

1. $(d * (d * y)) * y$. In this case $d * y$ can be x with the property as $d * x = y$, or $d * y = y$. In the first case we have $d * y = x$, and $d * x = y$ where $y * y = 0$. In the second case, $d * y = y$, and finally $y * y = 0$.

2. $(x * (x * d)) * d$. Here $x * d = 0$, $x * 0 = x$, and finally $x* = 0$.

Q.E.D.

Remark 3.6. We notice that in the previous extension, 0 is placed on the column of the element d, and on the line of the element d there are only elements with the property of involution, or auto involution, that is: $d * (d * x) = x$, respectively $d * x = x$.

Definition 3.7. We will call the extension defined previously the involution extension of a BCK-algebra of order n to an algebra of order $n + 1$.

Proposition 3.8. If a BCK-algebra $(X, *, 0)$ is commutative, the BCK-algebra resulting from the involution extension is also a commutative BCK-algebra.

Proof. We have the following cases:

1. $d * (d * y) = y * (y * d)$. In this case $y * d = 0$, $y * 0 = y$. $d * (d * y) = y$.

2. $x * (x * d) = d * (d * x)$. In this case, $x * d = 0$, and $x * 0 = x$. $d * (d * x)) = x$.

Q.E.D.

Proposition 3.9. Let $(X, *, 0)$ be a positive-implicative BCK-algebra. The BCK-algebra of order $n+1$ obtained by the extension through involutions is positive-implicative, if $(d*z)*(y*z) = (d*y)*z$.

Proof. We have the following cases:

1. $(d * z) * (y * z) = (d * y) * z$. It is true from the statement.
2. $(x * z) * (d * z) = (x * d) * z$. The right hand expression is 0. So it remains to be shown that $(x * z) * (d * z) = 0$. This expression is 0 from the definition of extension by involution, see the proof for axiom 1.
3. $(x * d) * (y * d) = (x * y) * d$. True, because the parentheses in front of the equals are 0, and after the equals, the expression is 0 from the definition of the element d.
4. $(x * d) * (d * d) = (x * d) * d$. The expression before equals is 0 from the definition of element d, and the one after equals is also 0.
5. $(d * d) * (y * d) = (d * y) * d$. The first parenthesis is 0, the entire expression becomes 0, and after all, everything is 0, from the definition of the element d.

Q.E.D.

Remark 3.10. Such a BCK-algebra exists:

*	0	1	2	3
0	0	0	0	0
1	1	0	1	0
2	2	2	0	0
3	3	2	1	0

Remark 3.11. In 3.9, we have the condition $(d*z)*(y*z) = (d*y)*z)$ which should be fulfilled by the resulting BCK-algebra to be a positive-implicative one. If we take $(d * y) = z)$ and $(d * z = y)$, we notice that the expression transforms in $y * (y * z) = z * z$, hence we have $y * (y * z) = 0$. In conclusion, if we have $(X, *, 0)$a positive-implicative BCK-algebra, an element $d \notin X$, satisfing the properties from the involution extension definition, for each pair of elements $(x, y) \in X$ with the property that $(d * x) = y$ and $d * y = x$, we must have $x * (x * y) = y * (y * x) = 0$

Remark 3.12. The following BCK-algebra does not obey the above rule:

*	0	1	2	3
0	0	0	0	0
1	1	0	0	0
2	2	2	0	0
3	3	2	1	0

$1 * (1 * 2) = 1$, $2 * (2 * 1) = 0$.
If we take the relation from the proposition: $(3 * 2) * (1 * 2) = (3 * 1) * 2$. $(3 * 1) = 2$, $(1 * 2) = 0$, $3 * 2 = 1$, so te expression becomes $1 * 0 = 2 * 2$, which means $1 = 0$ false.

Remark 3.13. By this process we can obtain bounded commutative BCK-algebras, starting from commutative BCK-algebras.

Remark 3.14. When passing from the order n, with n an even number, to the order $n+1$ at least one element on the line of the element d is a self-involution, that is $d * x = x$.

Proposition 3.15. Let $(X, *, 0)$ be a positive-implicative BCK-algebra of order n, with n an even number. If we want to extend this BCK-algebra to a BCK-algebra of order $(n + 1)$, using the involution extension, it must be a bounded BCK-algebra.

Proof. In 3.14 we have noticed that going from n, an even number, to $(n+1)$ trough involution extension, gives us a self-involution element on the last row, let it be y. It has the property that $(d*y) = y$. From the proof of the involution extension, we have seen that the following relation should be fulfilled: $(x*y)*(d*y) = 0$, for all $x, y \in X$. But there is an y with the property $d*y = y$, so, for this element, the expression reduces to $(x*y)*y = 0$. In any positive-implicative BCK-algebra, we have the relation $(x*y)*y = (x*y)$ [see [6]]. Therefore $(x*y) = 0$ for y with the property $d*y = y$. Hence the source algebra should be bounded with y the greatest element.

Q.E.D.

Remark 3.16. As we see, the involution extension does not play nicely with the positive-implicative property. In some cases, it makes this property to vanish, in other cases, we can not even apply this extension to positive-implicative BCK-algebras. On the other hand, it plays very nicely with the commutative property.

We further define a construction process, which is a particular case of this extension.

Corollary 3.17. Suppose we have a BCK-algebra X of order n, whose elements are numbered in the order of their appearance in the operations table from top-to down, 0 to $n-1$. We introduce a new element that we note with n, which respects the properties of the element d from the definition of extension by involution, with the following specific configuration of the table:

1. On the column of element n, all elements are 0;

2. On the line of the element n, all the elements are placed in reverse order from the one in the table. That is, the line looks like this: $n, (n-1), (n-2), ..., (n-(n-2)), 1, 0$

Remark 3.18. From the construction, we notice that the elements are involutions, that is: $n*0 = n$, and $n*n = 0$, $(n*1 = (n-1)$, and $n*(n-1) = 1$. In the case of passing from an order n even, to an order $n+1$ odd, the element in the middle of the line will be an auto involution IE: $n*([(n/2)]+1) = [(n/2)]+1)$.

Remark 3.19. It can be observed in some examples, that BCK-algebras obtained by extension through involutions, are isomorphic with BCK-algebras obtained by the process presented in 3.17. For example, we take the following two BCK-algebras:

$*$	0	1	2	3	4
0	0	0	0	0	0
1	1	0	0	0	0
2	2	2	0	3	0
3	3	3	0	0	0
4	4	2	1	3	0

$*$	0	1	2	3	4
0	0	0	0	0	0
1	1	0	0	0	0
2	2	2	0	0	0
3	3	3	2	0	0
4	4	3	2	1	0

The two algebras are isomorphic by changing 3 to 2, through an isomorphism, leaving the other elements in place. It remains to be studied how we can extend this example to a general result.

Remark 3.20. It also remains to be studied if there are ways to detect those BCK-algebras to which this extension can be applied, using only elements or properties of the source algebra, without involving the d element. Thus, given a BCK-algebra of order n, we can say whether or not this extension can be applied without using the element d.

3.2 Structure of the BCK-algebras class

In the following, we give some interesting results about the structure of the class of BCK-algebras and the importance of this structure. We introduce the notion of BCK-trees and their properties. First of all, we should pay close attention to how BCK-algebras can be build directly.

3.2.1 Construction of BCK-algebras

The axiomatic system generates some constraints on the operation table as follows:

1. On the 0 element line, there are only 0 values because $0 * x = 0$; from axiom 5;
2. Diagonally we only have 0 according to $x * x = 0$; from axiom 3
3. On the column of the 0 element, each element has its own value because $x * 0 = x$; from relation 9
4. We can not have 0 at the same time on position (i, j) and on position (j, i) with $j \neq i$ because $x * y = 0, y * x = 0 \rightarrow x = y$; from axiom 4
5. If the algebra is bounded, in the 1 column, the values are 0, because $x * 1 = 0$.

Basically, starting from these simple rules, you can build BCK-algebras and calculate the number of algebras of a particular order. So we find that we have a single algebra of order 2 that is bounded and implicative.

Example 3.21. Let X = 0, a.

$*$	0	a
0	0	0
a	a	0

For order 3, applying the axioms we get the following skeleton:

$*$	0	a	b
0	0	0	0
a	a	0	x
b	b	x	0

where we noted with x missing elements that must be completed. Note that the simplest rules at a 3 per 3 table leaves only two available positions, with a constraint however, we cannot put 0 on both positions where we have x. We obtain the following possible algebras:

Alg 1.

$*$	0	a	b
0	0	0	0
a	a	0	0
b	b	a	0

Alg 2.

*	0	a	b
0	0	0	0
a	a	0	0
b	b	b	0

Alg 3.

*	0	a	b
0	0	0	0
a	a	0	a
b	b	0	0

Alg 4.

*	0	a	b
0	0	0	0
a	a	0	b
b	b	0	0

Alg 5.

*	0	a	b
0	0	0	0
a	a	0	a
b	b	a	0

Alg 6.

*	0	a	b
0	0	0	0
a	a	0	b
b	b	b	0

Alg 7.

*	0	a	b
0	0	0	0
a	a	0	a
b	b	b	0

Alg 8.

*	0	a	b
0	0	0	0
a	a	0	b
b	b	a	0

Considering that we used the axioms 3, 4, 5 and 9 in the construction of the table and using the result in 2.3, we must check only the axiom 1. By calculations, the first 4 algebras are BCK-algebras, isomorphic two by two, 1 with 4 and 2 with 3. The algebra in 7 is a BCK-algebra, the rest being neither BCK-algebra nor BCI. We have 3 BCK-algebras of the order 3, unique to an isomorphism. For the study, we will retain as algebra representations 1, which is bounded, commutative but not positive-implicative, 2, bounded, positive-implicative but non-commutative and 7 implicative, therefore positive-implicative and commutative.

3.2.2 Structure of the operation table of a BCK-algebra

In 3.2.1 we have seen that the axiomatic system places certain constraints on the way the table of operations of a BCK-algebra looks. Starting from these constraints, we can construct BCK-algebra templates, templates that must then be filled in with the missing elements. For the order 3, the template looks like this:

$*$	0	a	b
0	0	0	0
A	A	0	x
b	b	x	0

Assuming we have an algebra of order 3 and we want to obtain, by extension, an algebra of order 4, we have the following template:

$*$	0	a	b	c
0	0	0	0	0
A	A	0	x	xx
b	b	x	0	xx
C	C	xx	xx	0

where, through x, we noted the unknown elements of the order 3, and through xx we noted the new elements introduced by the extension to the order 4. Such a template can be extended to the order n. We notice that if we have an algebra of order $n-1$ and we want to extend it to the order n, the axiomatic system leaves us, on the last column, $n-2$ elements to complete, and on the last line we have also $n-2$ elements, because $x*0=x$, $x*x=0$, $0*x=0$. In other words, the axiomatic system of BCK-algebras always fixes the corners of the algebra operating table. Each of the $n-2$ positions left on the last line and the last column can be filled with elements from 0 to $n-1$, that is n possibilities. By computation, the total number of possibilities for combining elements is $n^{(n-2)}$ for the line and $n^{(n-2)}$ for the column. This leads to a total number of possible combinations $(n^{(n-2)})^2$. From this total number, we must subtract, however, the constraints placed by the axiom according to which $x*y=0$ and $y*x=0 \to x=y$. In other words, in the template we build, we cannot put 0 at the same time on the corresponding positions, so if $x*y=0$, then the $y*x$ must be different from 0. The situations given by this axiom are $\sum_{i=1}^{n-2} C_i^{n-2} = 2^{(n-2)}-1$. Through these calculations, we demonstrated the following result:

Proposition 3.22. The total number of tables among which we can find a BCK-algebra when making the transition from the order $n-1$ to the order n, is $(n^{(n-2)})^2-(2^{(n-2)}-1)$.

By this result, we can calculate, for a certain order n, what is the number of tables among which we can find BCK-algebras, considering that the previous formula applies to any order.

In [6] Theorem 2.6 it is shown that any BCK-algebra of order n contains a BCK algebra of order $n-1$. We can therefore always see a BCK algebra of order n, as an extension of a BCK algebra of order $n-1$, obtained by bordering the operations table with a line and with a column. We get:

Proposition 3.23. The total number of n order templates to be checked when identifying n order BCK-algebras is:
$\prod_{i=2}^{n} (i^{(i-2)})^2-(2^{(i-2)}-1)$.

With this result, we observe that from very small orders it is difficult to find BCK-algebras, both by direct calculation and by the process of order-to-order extensions. Over this previously calculated number, if we consider an algorithm for verifying the axioms of BCK-algebras, we observe that for each such generated table, we have to perform n^3 operations to verify that the generated table is a BCK-algebra. So a legitimate question can be asked: Can we get a better algorithm for generating BCK-algebras? Can we somehow get BCK-algebras with certain properties in a faster way?

3.2.3 BCK-trees

In [6], Theorem 2.6 it is shown that any BCK-algebra of order n contains a BCK algebra of order $n-1$. Moreover, it is stated that a commutative algebra contains only commutative algebras. This perspective on BCK-algebras has always been seen from the top-down direction, from larger algebras to smaller algebras. By changing the perspective in reverse and looking from the bottom up, we can present the following conclusions, contained in the following results.

Proposition 3.24. Any BCK-algebra of order n, $n \geq 3$ contains at least one subalgebra of order 3

Proof. According to [6], theorem 2.6, every BCK-algebra of order n contains a BCK algebra of order $n-1$. Let $(X, *, 0)$ be a BCK-algebra of order n and $Y, *, 0)$, $Y \subset X$ a subalgebra in X of order $n-1$. Then Y contains a subalgebra of order $n-2$. Therefore, a chain of subalgebras is formed that goes down to the order of 3. Q.E.D.

Proposition 3.25. If a BCK-algebra contains a non-commutative subalgebra, it is non-commutative.

Proof. It emerges immediately from the definition of commutativity, given that it must be valid for any two elements x, y that belong to algebra. Q.E.D.

In fact, the proposition regarding commutativity is valid for any of the properties of BCK-algebras that refer to all elements of algebra. Thus, we can say:

Proposition 3.26. If a BCK-algebra contains a subalgebra that is not positive-implicative, it is not positive-implicative.

Corollary 3.27. If a BCK-algebra of order n contains an algebra of order 3 which is not commutative or positive-implicative, then it is not commutative or positive-implicative.

Proof. It emerges immediately from the propositions 3.24, 3.25, 3.26 Q.E.D.

In [6], Theorem 2.7, the exact number of subalgebras of a given order that a BCK-algebra of order n has is calculated. Thus, we can see that a BCK-algebra of order n contains many BCK-algebras of order 3.

Remark 3.28. Using the bottom-up perspective, we see that we can construct higher-order BCK-algebras starting from 3 algebra. And if the 3 algebra from which we start doesn't have a certain property, any higher-order BCK-algebra will not have that property.

Thus, starting from the BCK-algebra from 2, only higher non-commutative algebras can be obtained, because this is non-commutative. On the other hand, starting from this algebra, we are likely to find positive-implicative BCK-algebras. Starting from the BCK-algebra from 1, only higher-order algebras that are not positive-implicative can be obtained, because this is not positive-implicative, but is possible to find commutative algebras. Noting that the algebras in 1 and 2 are not implicative, it follows that any higher-order algebra chain starting from them will be non-implicative. The algebra from 7 is implicative, so commutative and positive-implicative. If we start from this, we have chances to obtain higher-order, implicative BCK algebras. Given that we have 3 algebras of order 3, unique to an isomorphism, it follows that we have 3 possible starting points on which to build higher-order BCK-algebras. Thus, true BCK-algebra trees are created, starting from the 3 types of BCK-algebras of the order of 3, trees which we will call BCK-trees.

Definition 3.29. The set of BCK-algebras, obtained by successive extensions from a lower order to a higher order, starting from a BCK-algebra of order 3 isomorphic with one of the algebras from 1, 2, 7, is called a BCK-tree.

Remark 3.30. BCK-trees are not independent, they intersect where, a higher-order algebra contains BCK-algebras of order 3 isomorphic with 2 or even all 3 types of algebra of order 3 from 1, 2 and 7, forming a true BCK-algebra graph. At these intersection points of the BCK-trees, the higher-order algebras, resulting from the extension, lose properties. They take the set of common properties held by the 3 order algebras they contain.

Remark 3.31. The intersection of the BCK-trees is made by means of BCK-algebras of the same order which are isomorphic between them.

For example, if we take the algebra from 7 and apply the Iseki extension, we get the following algebra:

$*$	0	1	2	3
0	0	0	0	0
1	1	0	1	0
2	2	2	0	0
3	3	3	3	0

By applying the Iseki extension, the resulting 4 order BCK-algebra contains an 3 order BCK-algebra isomorphic with the algebra of 2, losing the commutativity. Therefore, this algebra should be found in the BCK-tree generated by 7, the one we started from, but also in the tree generated by 2, the one with which it intersected. Given these things, we should be able, starting from the BCK-algebra of 2, to obtain the same BCK-algebra of the order of 4, or one isomorphic with it. Using the computer, we generated all BCK-algebras of order ≤ 5, and on the order of 4, starting from the algebra of 2, we found the following algebra:

$*$	0	1	2	3
0	0	0	0	0
1	1	0	0	1
2	2	2	0	2
3	3	3	0	0

This is obviously isomorphic with the previous algebra, through the application that switches the elements 2 and 3 between them, leaving in place the elements 0 and 1. We can now issue an important observation as a conclusion of this subsection:

Remark 3.32. The commutativity and positive-implicative properties of a BCK-algebra n with $n \geq 4$ depend on the properties of the BCK-algebras of order 3 that it contains as subalgebras.

3.2.4 The importance of BCK-trees

Regarding this new structure, we observe a similarity with the Cayley Dickson process from algebras built over a field, a process by which, at each step of transition from one dimension to another, certain properties are lost. This structure is useful to be studied, because it shows us where we can go to get algebras with certain properties. As we have seen, starting from one of the 3 order BCK-algebra types, a certain property may not exist. The importance of this structure is even greater if we consider the equivalence relations between the BCK-algebras class and other logical algebra classes. We know that there are links between Hilbert algebras and positive-implicative

BCK-algebras, [see [1]], between MV-algebras and bounded commutative BCK-algebras [see [8]], as well as between Boolean algebras and implicative bounded BCK-algebras [see [6]].

In this sense, if we want to find Boolean algebras, among the BCK-algebras, they will have to be searched only by extensions of the BCK-algebra of order 3 from the 7, because this is the only BCK-tree that contains implicative bounded BCK-algebras. Knowing that Boolean algebras have the order 2^n, we can find Boole algebras among BCK-algebras in the associated tree of the algebra from 7, taking the algebras which have the order 2^n. Specifically, we can generate them using an extension from a BCK-algebra of the order $2^n - 1$, positive-implicative and commutative, from that tree, using such an extension that gives us a bounded algebra and does not destroy the commutativity and positive-implicative property. The only extension to generate bounded BCK-algebras known is the Iseki extension, but, as we have seen, it destroys commutativity so it cannot be used for this purpose. In this paper, we proposed the involution extension as a mechanism to obtain bounded commutative BCK-algebras and, in some circumstances, bounded implicative BCK-algebras.

Regarding MV-algebras, given the equivalence relation with bounded and commutative BCK-algebras, it is clear that they will not be searched in the tree associated with the 2 algebra, because this algebra is not commutative. So we have the other two trees. We have noticed that in the tree associated with the algebra in 7 we have implicative algebras, and some of these can be bounded. Moreover, we have shown, by equivalence with Boolean algebras, that there exist algebras having the order 2^n in this tree that are bounded and implicative. In fact, we get a stronger result:

Proposition 3.33. Any bounded BCK-algebra with other order than 2^n, is not implicative.

However, bounded BCK-algebras, in this tree, are also found at other orders than 2^n, and even there are algebras of order 2^n that are not implicative. Indeed, it is sufficient to apply the Iseki extension to a BCK-algebra of order $2^n - 1$ in this tree, to obtain such a bounded non-implicative BCK-algebra of order 2^n. The bottom line is that for these bounded algebras with a different order than 2^n, one of the properties disappears, either the commutativity or the positive-implicative property. This means that they are at a point of intersection of this tree generated by the algebra of 7 with one of the other two trees. If the intersection is made with the tree generated by the algebra in the 2, the commutativity is lost, so this path does not help us. We can therefore conclude that MV-algebras are found either in the tree generated by the algebra in 1, or at the points of intersection of this tree with the tree generated by the algebra in 7, the algebra in 1 being itself a commutative bounded BCK-algebra. Using this assumptions and the fact that Boolean algebras class is a proper subclass of MV-algebras class, we can derive the following observation:

Remark 3.34. The difference between MV-algebras and Boolean Algebras in therm of BCK-algebras Properties is the positive-implicative property. This property is lost by MV algebras.

Finally, regarding the equivalence relation between Hilbert algebras and positive-implicative BCK-algebras, we can find these algebras in the trees generated by the algebras of 2 and 7, these being trees that contain positive-implicative algebras.

4 Conclusions

In this paper, we presented two new ways for extending BCK-algebras and the effect of these new extensions on BCK-algebras properties. They should be further investigated, to see how they are related with other properties, like the property to be with the condition(s.). In particular, The Involution extension could be further investigated to define exactly the class of BCK-algebras to which it can be applied directly, without checking the outside extension element. Also, we structured the class of BCK-algebras by introducing the BCK-tree concept. BCK-trees separate subclasses of BCK-algebras, based on their properties, so we have the tree of commutative BCK-algebras, the tree of positive-implicative BCK-algebras and the tree of Implicative BCK-algebras. This structure shows also the relation between BCK-algebras and other classes of logical algebras. Moreover, based on this, we build some conclusions about implicative BCK-algebras, conclusions that can be more developed in future research.

References

[1] W.A. Dudek, *On embedding Hilbert algebras in BCK-algebras*, Mathematica Moravica **3** (1999) 25–28.

[2] W.A. Dudek, *Remarks on the axioms system for BCI-algebras*, *Prace Naukowe WSP w Czestochowie* **2** (1996) 46–61.

[3] C. Flaut, *BCK-algebras arising from block-codes*, Journal of Intelligent & Fuzzy Systems **28** (2015) 1829–1833.

[4] Yasuyuki Imai and Kiyoshi Iseki, *On axiom systems of propositional calculi, XIV*,Proceedings of the Japan Academy **42** (1966).

[5] Y. B. Jun and S. Z. Song, *Codes based on BCK-algebras*, Inform. Sciences. **181** (2011) 5102–5109.

[6] J. Meng and Y. B. Jun, *BCK-Algebras*, Kyung Moon Sa Co., Seoul, (1994).

[7] S. R. Mukkamala, *A Course in BE-algebras*, Springer Nature Singapore Pte Ltd., Singapore, (2018).

[8] C. Flaut and R. Vasile, *Wajsberg algebras arising from binary block-codes*, Soft computing **24** (2020) 6047–6058.

DOI 10.2478/9788395793882-012

On Lorentz transformations with elliptic biquaternions

Zülal Derin[1] and Mehmet Ali Güngör[2]

[1,2] Department of Mathematics, Faculty of Arts and Sciences, Sakarya, Turkey
E-mail: zulal.derin1@ogr.sakarya.edu.tr[1], agungor@sakarya.edu.tr[2]

Abstract

In this study, the Lorentz transformations which are in accordant with special relativity have been examined for the first time with elliptic biquaternions. Since the elliptic biquaternions contain the complex structure, it is beneficial to examine the Lorentz transformations which can be the building blocks of relativistic physics via elliptic biquaternions. Therefore, as a result of relativistic transformation relation, it has been seen that the Lorentz transformations can be expressed with elliptic biquaternions and some special results have been given. In addition, matrix representations of obtained mathematical expressions are given. Thanks to the matrix representations of elliptical biquaternions, the property of commutativeness which is not valid for elliptic biquaternions has been eliminated and these representations provide a convenience for relativistic transformation relation. In this context, the presented method in this article is very useful.

2010 Mathematics Subject Classification. **11R52** 00A79
Keywords. elliptic biquaternions, Lorentz transformations.

1 Introduction

Quaternions were discovered in 1843 during the studies of Irish mathematician William Rowan Hamilton while extending the complex numbers into three-dimensional space. From that day forward, the quaternions were used in physics by E. Schrödinger, W. Heisenberg, P. A. M. Dirac, M. Born and many other famous physicists between 1927 and 1932 in parallel with the developments of quantum mechanics. Real quaternions have applications in many fields such as differential geometry, motion geometry, quantum mechanics and real quaternions sentence can be represented as

$$\mathbb{H} = \{\mathbf{Q} = a_0 + a_1 i + a_2 j + a_3 k : a_0, a_1, a_2, a_3 \in \mathbb{R}\}$$

where the quaternion unit bases $1, i, j$ and k satisfies the multiplication laws as follows:

$$i^2 = j^2 = k^2 = -1,\ ij = -ji = k,\ jk = -kj = i,\ ki = -ik = j$$

[1]. Also, the quaternionic units i, j, k are taken as base vectors of the $3-$ dimensional vector space, so that a quaternion $\mathbf{Q}$ can be written as a linear combination of a scalar and a spatial vector [2]. In 1853 W. R. Hamilton defined the sentence biquaternions (complex quaternions). The quaternionic units i, j, k are the same with the units in the real quaternions, this sentence is defined by

$$\mathbb{H}_{\mathbb{C}} = \{\mathbf{Q} = W + Xi + Yj + Zk : W, X, Y, Z \in \mathbb{C}\}$$

[3]. Complex quaternions have many applications in mathematics and physics. The applications of complex quaternions in physics have been mostly in the field of general and special relativity, relativistic mechanics, electromagnetism and quantum mechanics [4-9]. One of these is the expression

Tbilisi Mathematical Journal Special Issue (IECMSA-2019), pp. 125–144.
Tbilisi Centre for Mathematical Sciences.

Received by the editors: 19 October 2019.
Accepted for publication: 20 April 2020.

of the particle mechanical equations, the conservation of 4-momentum formulas and the equations of electromagnetism by means of complex quaternions [10]. It is seen that many researchers have searched out the studies using the formulation of complex quaternions on Lorentz transformations and Maxwell equations [11-15]. Afterwards, it is seen that the complex quaternion algebra and the whole 2×2 complex matrix algebra are isomorphic to each other. This isomorphism is defined as

$$\psi : \mathbb{H}_{\mathbb{C}} \to M_2(\mathbb{C}) \ , \ \psi(a_0 + a_1 i + a_2 j + a_3 k) = \begin{bmatrix} a_0 + a_1 i & -a_2 - a_3 i \\ a_2 - a_3 i & a_0 - a_1 i \end{bmatrix}$$

[16-19]. By means of this isomorphism, many researchers have done studies. For instance, the matrix expressions of Lorentz transformations with complex quaternions studied by S. Demir [20]. On the other hand, the isomorphism for the 2x2 elliptical matrix representation of elliptical biquaternions is defined by Özen et al. [21] as follows:

$$\sigma : \mathbb{HC}_p \to M_2(\mathbb{C}_p), \mathbf{Q} = A_0 + A_1 i + A_2 j + A_3 k \to \sigma(\mathbf{Q}) = \begin{bmatrix} A_0 + \frac{1}{\sqrt{|p|}} I A_1 & -A_2 - \frac{1}{\sqrt{|p|}} I A_3 \\ A_2 - \frac{1}{\sqrt{|p|}} I A_3 & A_0 - \frac{1}{\sqrt{|p|}} I A_1 \end{bmatrix} \quad (1.1)$$

where $\mathbf{Q}$ and $\mathbf{R}$ are any elliptic biquaternions, the function σ is a linear isomorphism bijection and surjection which provides properties $\sigma(\mathbf{Q}+\mathbf{R}) = \sigma(\mathbf{Q}) + \sigma(\mathbf{R})$ $\sigma(\mathbf{QR}) = \sigma(\mathbf{Q})\sigma(\mathbf{R})$. $\mathbb{HC}_p$ and $M_2(\mathbb{C}_p)$ denote the set of elliptic biquaternions and the set of elliptic matrices of type 2×2 respectively. In this study, we investigate Lorentz transformations with elliptic biquaternions through the isomorphism described above. We define special matrices with elliptic biquaternions that we obtained under this isomorphism and using these special matrices we give the 4×4 elliptic and 8×8 type real matrices. Also, by means of matrix representations corresponding to the left Hamiltonian operator, the problem of the non-commutative property of the elliptic biquaternions in the algebraic structure has disappeared. Then we expressed the elliptic biquaternion $\mathbf{R}$ that relates space and time with equation 4.1. In Theorem 4.1, we give the elliptical matrix representation of this expression. In this way, the space and time components of the elliptic biquaternion $\mathbf{R}'$ obtained as a result of the relativistic transformation relation can be easily seen. In this expression, the real component of the elliptic biquaternion $\mathbf{R}' = ct + I\boldsymbol{r}$ denotes time and its the imaginer component denotes space. Note here that the imaginer component $I = \sqrt{|p|}$ denotes the space. Moreover, by taking $I^2 = p = -1$ since $I^2 = p < 0$, we also showed that elliptic biquaternions include complex cases.

2 Complex numbers and elliptic biquaternions

2.1 $p-$Complex numbers

The pairs (x, y) whose elements are real numbers are called complex numbers. The first persons to benefit from complex numbers are G. Cardan and R. Bombelli. On the other hand, the generalized complex numbers are separable to three as ordinary, dual and double complex numbers. The complex numbers defined as $i^2 = -1$ in case of imaginary unit i. It is the case the imaginary unit i the natural complex numbers are $i^2 = -1$. The English geometrician W. Clifford developed double complex numbers for the case of $i^2 = 1$. The German geometrician E. Study then carried out the studies on kinematics and line geometry, introducing different theorems and in the case of $i^2 = 0$ the dual numbers are obtained. I. Yaglom expressed that the ordinary, dual and double numbers are special members of the two-parameter family of the complex number system and he defined generalized complex numbers as

$$z = x + iy \ (x, y \in \mathbb{R}) \ , \ i^2 = iq + p \ (q, p \in \mathbb{R})$$

[22]. Harkin et al. have expressed that the generalized complex numbers are isomorphic to other number systems under consideration $i^2 = p$ and $q = 0$ in $i^2 = iq + p$. Then $p-$complex numbers system $\mathbb{C}_p$ is defined as

$$\mathbb{C}_p = \left\{x + iy : \ x, y \in \mathbb{R}, \ i^2 = p\right\}.$$

Also, $p-$ trigonometric functions are defined as

$$\cos_p\theta_p = \begin{cases} \cos\ (\theta_p\sqrt{|p|}) & , p < 0 \\ 1 & , p = 0 \\ \cosh(\theta_p\sqrt{p}) & , p > 0 \end{cases} \tag{2.1}$$

and

$$sin_p\theta_p = \begin{cases} \frac{1}{\sqrt{|p|}}sin\ (\theta_p\sqrt{|p|}) & , p < 0 \\ 1 & , p = 0 \\ \frac{1}{\sqrt{p}}sin(\theta_p\sqrt{p}) & , p > 0 \end{cases} \tag{2.2}$$

[23]. Also Harkin expressed in his this geometric study that generalized complex numbers and rotations can be applied to the theory of special relativity in physics. In another studies on $p-$complex numbers, the generalized Steiner formula and the Holditch theorem were presented by T. Erişir and M. A. Güngör [24]-[25]. The exponential function for e^{φ_p} an elliptic number $\varphi_p = x + Iy \in \mathbb{C}_p$ in the set of elliptic numbers $\mathbb{C}_p$ is known as a $p-$ analytic function defined in the form of $e^{\varphi_p} = e^{x+Iy} = e^x e^{Iy} = e^x(\cos_p y + I\sin_p y)$. From here, the $p-$ trigonometric functions can be obtained as follows:

$$\begin{cases} \cos_p y = \frac{e^{Iy}+e^{-Iy}}{2}, \ \sin_p y = \frac{e^{Iy}-e^{-Iy}}{2I} & \text{for} \quad x = 0 \\ \cos_p(Iy) = \frac{e^{py}+e^{-py}}{2} = \cosh(py) & \text{for} \quad y = Iy \\ \sin_p(Iy) = \frac{e^{py}-e^{-py}}{2I} = \frac{I}{p}\sinh(py) & \text{for} \quad y = Iy \end{cases} \tag{2.3}$$

On the other hand, for the elliptic complex variable $p-$ cosine and $p-$ sine functions also provide the identity ${\cos_p}^2\varphi_p + |p|\,{\sin_p}^2\varphi_p = 1$ where $\varphi_p = x + Iy \in \mathbb{C}_p$ [26].

2.2 Elliptic biquaternions

The set of elliptic biquaternions is given in the cartesian form as follows:

$$\mathbb{HC}_p = \left\{\mathbf{Q} = \mathbf{q} + I\mathbf{q}' = A_0e_0 + A_1e_1 + A_2e_2 + A_3e_3 \,:\, A_0, A_1, A_2, A_3 \in \mathbb{C}_p, I^2 = p < 0\right\}$$

where the numbers $A_i = q_i + Iq_i'$, $0 \leqslant i \leqslant 3$ state elliptic numbers. The unit bases of a biquaternion provide the product rule given in Table 1 [1]. Let $\mathbf{Q} = A_0e_0 + A_1e_1 + A_2e_2 + A_3e_3 \in \mathbb{HC}_p$ be an elliptic biquaternion providing the product rule in Table 1, then $\mathbf{Q}$ is defined as pure elliptic biquaternions provided that $A_0 = 0$ [27].

Any elliptic biquaternion $\mathbf{P}$ is expressed with the scalar (B_0) and vectorial components ($B_1e_1 + B_2e_2 + B_3e_3$) in the following form

$$\mathbf{P} = S(\mathbf{P}) + V(\mathbf{P}) = B_0 + \boldsymbol{P} \tag{2.4}$$

TABLE 1. Multiplication scheme of the quaternionic units.

$\times$	e_0	e_1	e_2	e_3
e_0	e_0	e_1	e_2	e_3
e_1	e_1	$-e_0$	e_3	$-e_2$
e_2	e_2	$-e_3$	$-e_0$	e_1
e_3	e_3	e_2	$-e_1$	$-e_0$

The sum and quaternionic product of any two elliptic biquaternions $\mathbf{Q} = A_0e_0+A_1e_1+A_2e_2+A_3e_3$ and $\mathbf{P} = B_0e_0 + B_1e_1 + B_2e_2 + B_3e_3$ are defined as

$$\mathbf{P} + \mathbf{Q} = (B_0e_0 + B_1e_1 + B_2e_2 + B_3e_3) + (A_0e_0 + A_1e_1 + A_2e_2 + A_3e_3)$$

$$\begin{aligned}\mathbf{P} + \mathbf{Q} &= (B_0 + A_0) + (B_1 + A_1)e_1 + (B_2 + A_2)e_2 + (B_3 + A_3)e_3\\ \mathbf{PQ} &= (B_0e_0 + B_1e_1 + B_2e_2 + B_3e_3)(A_0e_0 + A_1e_1 + A_2e_2 + A_3e_3)\\ \mathbf{PQ} &= (B_0A_0 - B_1A_1 - B_2A_2 - B_3A_3) + (B_1A_0 + B_0A_1 + B_3A_2 - B_2A_3)e_1\\ &\quad +(B_2A_0 - B_3A_1 + B_0A_2 + B_1A_3)e_2 + (B_3A_0 + B_2A_1 - B_1A_2 + B_0A_3)e_3\ .\end{aligned}$$

In other words, the quaternionic product of two elliptic biquaternions can also be given, in the form of

$$\mathbf{PQ} = (B_0 + \boldsymbol{B})(A_0 + \boldsymbol{A}) = B_0A_0 + B_0\boldsymbol{A} + A_0\boldsymbol{B} - \langle \boldsymbol{B}, \boldsymbol{A}\rangle + \boldsymbol{B} \wedge \boldsymbol{A} \tag{2.5}$$

where $\langle , \rangle$ and $\wedge$ indicate inner product and vector product in three-dimensional space. The elliptic biquaternion $\mathbf{Q} = A_0e_0 + A_1e_1 + A_2e_2 + A_3e_3 = S(\mathbf{Q}) + V(\mathbf{Q}) = \mathbf{q} + I\mathbf{q}'$ has three different conjugates given as follows:

$$\begin{aligned}\bar{\mathbf{Q}} &= A_0e_0 - A_1e_1 - A_2e_2 - A_3e_3 = S(\mathbf{Q}) - V(\mathbf{Q})(\text{quaternionic conjugate})\\ \mathbf{Q}^* &= (A_0)^*e_0 + (A_1)^*e_1 + (A_2)^*e_2 + (A_3)^*e_3\\ &= (S(\mathbf{Q}))^* + (V(\mathbf{Q}))^* = \mathbf{q} - I\mathbf{q}'(\text{complex conjugate})\\ \mathbf{Q}^\dagger &= (\bar{\mathbf{Q}})^* = \overline{(\mathbf{Q}^*)} = (A_0)^*e_0 - (A_1)^*e_1 - (A_2)^*e_2 - (A_3)^*e_3\\ &= (S(\mathbf{Q}))^* - (V(\mathbf{Q}))^* = \bar{\mathbf{q}} - I\bar{\mathbf{q}}' \text{ (total conjugate)}\end{aligned}$$

where $\bar{\mathbf{q}}$ and $\bar{\mathbf{q}}'$ indicate the biquaternionic conjugates of real quaternions $\mathbf{q}$ and $\mathbf{q}'$ respectively. On the other hand, the inner product of these elliptic biquaternions is defined in the following way:

$$\langle \mathbf{P}, \mathbf{Q}\rangle_Q = \frac{1}{2}(\bar{\mathbf{P}}\mathbf{Q} + \bar{\mathbf{Q}}\mathbf{P}) = \frac{1}{2}(\mathbf{P}\bar{\mathbf{Q}} + \mathbf{Q}\bar{\mathbf{P}})$$

where the symbol $\langle\rangle_Q$ indicates a quaternionic inner product. Using this inner product semi-norm of $\mathbf{Q}$ is expressed as follows:

$$N(\mathbf{Q}) = \langle \mathbf{Q}, \mathbf{Q}\rangle_Q = \mathbf{Q}\bar{\mathbf{Q}} = \bar{\mathbf{Q}}\mathbf{Q} = {A_0}^2 + {A_1}^2 + {A_2}^2 + {A_3}^2$$

Also, since $A_i = q_i + Iq_i' \in \mathbb{C}_p$ $(0 \leqslant i \leqslant 3)$ are in the above equation, it is seen that $N(\mathbf{Q})$ can be equal to zero while $\mathbf{Q} \neq 0$. Accordingly, in the algebra of the elliptic biquaternions $\mathbb{HC}_p$, there are some elliptic biquaternions $\mathbf{Q}$ that satisfy the equality $\mathbf{Q}\bar{\mathbf{Q}}$ even if $\mathbf{Q} \neq 0$ and the algebra of the elliptic biquaternions $\mathbb{HC}_p$ contains zero divisors. Therefore, semi-norm is used instead of the norm in the elliptic biquaternion space $\mathbb{HC}_p$ in order to be appropriateness to the general literature [26]. Provided the semi-norm of the elliptic biquaternion $\mathbf{Q}$ is different from zero then it's inverse is given by $\mathbf{Q}^{-1} = \frac{\bar{\mathbf{Q}}}{N(\mathbf{Q})}$. Also, the module of this elliptic biquaternion $\mathbf{Q}$ is defined as

$$N(\mathbf{Q}) = \mathbf{Q}\bar{\mathbf{Q}} = \bar{\mathbf{Q}}\mathbf{Q} = |\mathbf{Q}|^2$$

and indicated by $|\mathbf{Q}|$. In the case of

$$N(\mathbf{Q}) = \langle \mathbf{Q}, \mathbf{Q} \rangle_Q = \mathbf{Q}\bar{\mathbf{Q}} = \bar{\mathbf{Q}}\mathbf{Q} = {A_0}^2 + {A_1}^2 + {A_2}^2 + {A_3}^2 = 1 \tag{2.6}$$

is satisfied, then the elliptic biquaternion $\mathbf{Q}$ is called unit elliptic biquaternions [26]. The set of all unit elliptic biquaternions is given by
$\eta = \{\mathbf{Q} \in \mathbb{HC}_p : N(\mathbf{Q}) = 1\}$ and the set of all pure elliptic biquaternions whose semi-norm is equal to the absolute value of p such as

$$\zeta = \{w = A_1e_1 + A_2e_2 + A_3e_3 : N(w) = {A_1}^2 + {A_2}^2 + {A_3}^2 = |p|, w^* = -w, w^2 = p\}.$$

Thus, for any elliptic biquaternion $\mathbf{Q}$ whose modulus never vanishes, this elliptic biquaternion satisfies the following equalities [26]

$$R = \left(\sqrt{{A_0}^2 + {A_1}^2 + {A_2}^2 + {A_3}^2}\right)_{\mathbb{C}_p}, \ \cos_p\varphi_p = \frac{A_0}{R}, \ \sin_p\varphi_p = \frac{\left(\sqrt{{A_1}^2 + {A_2}^2 + {A_3}^2}\right)_{\mathbb{C}_p}}{\sqrt{|p|}R}. \tag{2.7}$$

Accordingly, the following theorem can be given.

Theorem 2.1. Let $w \in \zeta$ be a pure elliptic biquaternion and $\varphi_p \in \mathbb{C}_p$ be an elliptical complex angle. In this case, the elliptic biquaternion can be written as $\cos_p\varphi_p + w\sin_p\varphi_p$ is a unit elliptic biquaternion [26].

Theorem 2.2. Let $\mathbf{Q} = A_0 + A_1e_1 + A_2e_2 + A_3e_3$ be the elliptic biquaternion whose modulus is non-zero. In this case, a pure elliptic biquaternion is derived from the vectorial part of such that

$$w_{\mathbf{Q}} = \sqrt{|p|}\frac{A_1e_1 + A_2e_2 + A_3e_3}{\left(\sqrt{{A_1}^2 + {A_2}^2 + {A_3}^2}\right)_{\mathbb{C}_p}} \tag{2.8}$$

is an element of the set ζ, i.e., $(w_{\mathbf{Q}})^2 = p$ [26].

The elliptic biquaternion $\mathbf{Q} = A_0 + A_1i + A_2j + A_3k$ in $p-$ trigonometric form depending on Theorem (2.1) and Theorem (2.2) can be written as

$$\mathbf{Q} = R(cos_p\varphi_p + w_{\mathbf{Q}}\sin_p\varphi_p). \tag{2.9}$$

Here the angle $\varphi_p \in \mathbb{C}_p$ is an elliptic complex angle in the form of $\varphi_p = x + Iy$ [26].

Theorem 2.3. If an elliptic biquaternion is given as $\mathbf{Q} = \cosh(p\frac{\theta_p}{2}) + \frac{1}{I}\hat{q}\sinh(p\frac{\theta_p}{2})$ then $\mathbf{Q}$ is a unit elliptic biquaternion

Proof. For a given velocity vector $\vec{v} = v_1 e_1 + v_2 e_2 + v_3 e_3$ and pure unit elliptic biquaternion $\hat{q} = \sqrt{|p|}\frac{v_1 e_1 + v_2 e_2 + v_3 e_3}{\sqrt{<\vec{v},\vec{v}>}}$, $\mathbf{Q} = \cosh(py) + \frac{1}{I}\hat{q}\sinh(py)$ can be written as unit elliptic biquaternion. By means of the equalities 2.3, 2.8 and 2.9 the conjugate of the elliptic biquaternion is taken as $\bar{\mathbf{Q}} = \cosh(py) - \frac{1}{I}\hat{q}\sinh(py)$. Here, if we give $y = \frac{\theta_p}{2}$, we can obtain as

$$\mathbf{Q}\bar{\mathbf{Q}} = N(\mathbf{Q}) = \cosh^2(p\frac{\theta_p}{2}) - \sinh^2(p\frac{\theta_p}{2}) = 1.$$

Q.E.D.

3 Matrix representations of elliptic biquaternions

In terms of more descriptive mathematical expressions, it is also the preferred method to give this mathematical expressions matrix representations. In this section, we study with matrices. It is possible to represent an elliptic biquaternion $\mathbf{Q}$ with 4×4 matrices. These matrices that similar to Pauli spin matrices can be defined with the help of isomorphism expressed in 1.1 as follows:

$$\begin{gathered}\sigma(e_0) = \sigma_0 = \begin{bmatrix} 1 & 0 \\ 0 & 1 \end{bmatrix}, \sigma(e_1) = \sigma_1 = \begin{bmatrix} \frac{I}{\sqrt{|p|}} & 0 \\ 0 & -\frac{I}{\sqrt{|p|}} \end{bmatrix} \\ \sigma(e_2) = \sigma_2 = \begin{bmatrix} 0 & -1 \\ 1 & 0 \end{bmatrix}, \sigma(e_3) = \sigma_3 = \begin{bmatrix} 0 & -\frac{I}{\sqrt{|p|}} \\ -\frac{I}{\sqrt{|p|}} & 0 \end{bmatrix}.\end{gathered} \tag{3.1}$$

These matrices satisfy the properties of

$$(\sigma(e_0))^2 = 1,\ (\sigma(e_1))^2 = (\sigma(e_2))^2 = (\sigma(e_3))^2 = -I_2.$$

In addition with the help of these special matrices, we define the matrices corresponding to the base elements e_0, e_1, e_2, e_3 in the real quaternions as follows:

$$\begin{gathered}\Gamma_0 = \begin{bmatrix} \sigma_0 & 0 \\ 0 & \sigma_0 \end{bmatrix}_{2\times 2} = \begin{bmatrix} 1 & 0 & 0 & 0 \\ 0 & 1 & 0 & 0 \\ 0 & 0 & 1 & 0 \\ 0 & 0 & 0 & 1 \end{bmatrix}, \Gamma_1 = \begin{bmatrix} \sigma_2 & 0 \\ 0 & -\sigma_2 \end{bmatrix} = \begin{bmatrix} 0 & -1 & 0 & 0 \\ 1 & 0 & 0 & 0 \\ 0 & 0 & 0 & 1 \\ 0 & 0 & -1 & 0 \end{bmatrix} \\ \Gamma_2 = \begin{bmatrix} 0 & -\sigma_0 \\ \sigma_0 & 0 \end{bmatrix} = \begin{bmatrix} 0 & 0 & -1 & 0 \\ 0 & 0 & 0 & -1 \\ 1 & 0 & 0 & 0 \\ 0 & 1 & 0 & 0 \end{bmatrix}, \Gamma_3 = \begin{bmatrix} 0 & \sigma_2 \\ \sigma_2 & 0 \end{bmatrix} = \begin{bmatrix} 0 & 0 & 0 & -1 \\ 0 & 0 & 1 & 0 \\ 0 & -1 & 0 & 0 \\ 1 & 0 & 0 & 0 \end{bmatrix}.\end{gathered} \tag{3.2}$$

These matrices satisfy the multiplication relations for basis of elliptic biquaternion as

$$\Gamma_0{}^2 = \Gamma_0 = I_4, \quad \Gamma_1{}^2 = \Gamma_2{}^2 = \Gamma_3{}^2 = -I_4, \quad \Gamma_j\Gamma_k = \delta_{jk}\Gamma_0 - \varepsilon_{jkl}\Gamma_l \tag{3.3}$$

where the expressions δ and ε denote Kronecker Delta and Levi-Civita symbols, respectively. Thus for an elliptic biquaternion $\mathbf{Q} \cong A_0\Gamma_0 + A_1\Gamma_1 + A_2\Gamma_2 + A_3\Gamma_3$ by using quaternionic bases in 3.2

the known left Hamiltonian matrix can be given by

$$\mathbb{H}^{-}(\mathbf{Q})=\begin{bmatrix} A_0 & -A_1 & -A_2 & -A_3 \\ A_1 & A_0 & A_3 & -A_2 \\ A_2 & -A_3 & A_0 & A_1 \\ A_3 & A_2 & -A_1 & A_0 \end{bmatrix} \tag{3.4}$$

On the other hand, the known right Hamiltonian matrix is as follows:

$$\mathbb{H}^{+}(\mathbf{Q})=\begin{bmatrix} A_0 & -A_1 & -A_2 & -A_3 \\ A_1 & A_0 & -A_3 & +A_2 \\ A_2 & +A_3 & A_0 & -A_1 \\ A_3 & -A_2 & +A_1 & A_0 \end{bmatrix}.$$

The number of elliptic biquaternions $\mathbf{Q}$ here is also; it is also possible to represent with the matrix 4×1 given by $\mathbf{Q}=[A_0,\boldsymbol{A}]^T=\begin{bmatrix} A_0 & A_1 & A_2 & A_3 \end{bmatrix}^T$.
The following theorem can be given for Hamiltonian matrices.

Theorem 3.1. Let two elliptic biquaternions be $\mathbf{Q}=A_0\Gamma_0+A_1\Gamma_1+A_2\Gamma_2+A_3\Gamma_3$ and $\mathbf{P}=B_0\Gamma_0+B_1\Gamma_1+B_2\Gamma_2+B_3\Gamma_3$ in algebra $\mathbb{HC}_p$. Accordingly, the following properties are provided:

$$\mathbf{Q}=\mathbf{R}\Leftrightarrow\mathbb{H}^{+}(\mathbf{Q})=\mathbb{H}^{+}(\mathbf{R})\Leftrightarrow\mathbb{H}^{-}(\mathbf{Q})=\mathbb{H}^{-}(\mathbf{R})$$

1. $\mathbb{H}^{+}(\mathbf{Q}+\mathbf{R})=\mathbb{H}^{+}(\mathbf{Q})+\mathbb{H}^{+}(\mathbf{R}),\quad \mathbb{H}^{-}(\mathbf{Q}+\mathbf{R})=\mathbb{H}^{-}(\mathbf{Q})+\mathbb{H}^{-}(\mathbf{R})$
2. $\mathbb{H}^{+}(\lambda\mathbf{Q})=\lambda\mathbb{H}^{+}(\mathbf{Q}),\quad \mathbb{H}^{-}(\lambda\mathbf{Q})=\lambda\mathbb{H}^{-}(\mathbf{Q})$
3. $\mathbb{H}^{+}(\mathbf{QR})=\mathbb{H}^{+}(\mathbf{Q})\mathbb{H}^{+}(\mathbf{R}),\quad \mathbb{H}^{-}(\mathbf{QR})=\mathbb{H}^{-}(\mathbf{R})\mathbb{H}^{-}(\mathbf{Q})$
4. $\mathbb{H}^{+}(\bar{\mathbf{Q}})=[\mathbb{H}^{+}(\mathbf{Q})]^T,\quad \mathbb{H}^{-}(\bar{\mathbf{Q}})=[\mathbb{H}^{-}(\mathbf{Q})]^T$
5. $\mathbb{H}^{+}(\mathbf{Q}^*)=\overline{\mathbb{H}^{+}(\mathbf{Q})},\quad \mathbb{H}^{-}(\mathbf{Q}^*)=\overline{\mathbb{H}^{-}(\mathbf{Q})}$
6. $\mathbb{H}^{+}(\mathbf{Q}^\dagger)=[\mathbb{H}^{+}(\mathbf{Q})]^*,\quad \mathbb{H}^{-}(\mathbf{Q}^\dagger)=[\mathbb{H}^{-}(\mathbf{Q})]^*$

where is an elliptical number [27].

On the other hand, by using the above properties it is possible to define the elliptic biquaternion multiplication given in equation 2.5 as follows:

$$\mathbb{H}^{-}(\mathbf{P})\mathbf{Q}=\begin{bmatrix} B_0 & -B_1 & -B_2 & -B_3 \\ B_1 & B_0 & B_3 & -B_2 \\ B_2 & -B_3 & B_0 & B_1 \\ B_3 & B_2 & -B_1 & B_0 \end{bmatrix}\begin{bmatrix} A_0 \\ A_1 \\ A_2 \\ A_3 \end{bmatrix}. \tag{3.5}$$

Now, let we the following antisymmetric matrices are define follow as:

$$\tilde{\boldsymbol{P}}=\begin{bmatrix} 0 & B_3 & -B_2 \\ -B_3 & 0 & B_1 \\ B_2 & -B_1 & 0 \end{bmatrix} \tag{3.6}$$

and

$$\tilde{\boldsymbol{Q}} = \begin{bmatrix} 0 & A_3 & -A_2 \\ -A_3 & 0 & A_1 \\ A_2 & -A_1 & 0 \end{bmatrix}. \tag{3.7}$$

In this case with the help of the defined matrix in 3.6 the expression 3.5 can be given as follows:

$$\mathbf{PQ} \cong \begin{bmatrix} B_0 & -\boldsymbol{B} \\ \boldsymbol{B^T} & B_0 I_3 + \tilde{\boldsymbol{P}} \end{bmatrix} \begin{bmatrix} A_0 \\ \boldsymbol{A} \end{bmatrix}. \tag{3.8}$$

Similarly using the right Hamiltonian matrix representation of multiplication the equality can be written as follows:

$$\mathbb{H}^{+}(\mathbf{Q})\mathbf{P} = \begin{bmatrix} A_0 & -A_1 & -A_2 & -A_3 \\ A_1 & A_0 & -A_3 & A_2 \\ A_2 & A_3 & A_0 & -A_1 \\ A_3 & -A_2 & A_1 & A_0 \end{bmatrix} \begin{bmatrix} B_0 \\ B_1 \\ B_2 \\ B_3 \end{bmatrix}. \tag{3.9}$$

On the other hand, considering the elliptic biquaternionic matrix given in 3.7, this multiplication matrix can be expressed as an elliptic matrix of size 2x2

$$\mathbf{PQ} = \begin{bmatrix} A_0 & -\boldsymbol{B} \\ \boldsymbol{B^T} & A_0 I_3 - \tilde{\boldsymbol{Q}} \end{bmatrix} \begin{bmatrix} B_0 \\ \boldsymbol{B} \end{bmatrix} \tag{3.10}$$

As is known, the product of two elliptic biquaternions has no commutative property. Therefore we define two elliptic biquaternions as follows:

$$\breve{\boldsymbol{P}} = \begin{bmatrix} B_0 & -\boldsymbol{B} \\ \boldsymbol{B^T} & B_0 I_3 + \tilde{\boldsymbol{P}} \end{bmatrix} \tag{3.11}$$

and

$$\breve{\mathbf{Q}} = \begin{bmatrix} A_0 & -\boldsymbol{A} \\ \boldsymbol{A^T} & A_0 I_3 - \tilde{\boldsymbol{Q}} \end{bmatrix} \tag{3.12}$$

$$\mathbb{H}^{-}(\mathbf{P})\mathbf{Q} \cong \breve{\mathbf{P}}\mathbf{Q}$$

and

$$\mathbb{H}^{+}(\mathbf{Q})\mathbf{P} \cong \breve{\mathbf{Q}}\mathbf{P}.$$

Hence, we can write the following equality

$$\mathbb{H}^{-}(\mathbf{P})\mathbf{Q} = \mathbb{H}^{+}(\mathbf{Q})\mathbf{P}$$

and

$$\breve{\mathbf{P}}\mathbf{Q} = \breve{\mathbf{Q}}\mathbf{P}. \tag{3.13}$$

Thus, the commutative property which is not valid for elliptic biquaternions can be easily obtained with the help of matrices. This equality will provide convenience for matrix representations of the relativistic transformation relation. It is also possible to express the elliptic biquaternion $\mathbf{Q} = \mathbf{q} + I\mathbf{q}'$ consisting of eight real components with the 8×8 real matrix using the expressions 3.2 defined for the base elements of the real quaternions.

Theorem 3.2. For the elliptical biquaternion $\mathbf{Q}$ there is the relation:
$\mathbb{H}^-(\mathbf{Q}) \cong (a_0 + T{a'}_0)\xi_0 + (a_1 + T{a'}_1)\xi_1 + (a_2 + T{a'}_2)\xi_2 + (a_3 + T{a'}_3)\xi_3$
$\mathbb{H}^-(\mathbf{Q}) \cong$

$$\begin{bmatrix} a_0 & -a_1 & -a_2 & -a_3 & \sqrt{|p|}a_0{}' & -\sqrt{|p|}a_1{}' & -\sqrt{|p|}a_2{}' & -\sqrt{|p|}a_3{}' \\ a_1 & a_0 & a_3 & -a_2 & \sqrt{|p|}a_1{}' & \sqrt{|p|}a_0{}' & \sqrt{|p|}a_3{}' & -\sqrt{|p|}a_2{}' \\ a_2 & -a_3 & a_0 & a_1 & \sqrt{|p|}a_2{}' & -\sqrt{|p|}a_3{}' & \sqrt{|p|}a_0{}' & \sqrt{|p|}a_1{}' \\ a_3 & a_2 & -a_1 & a_0 & \sqrt{|p|}a_3{}' & \sqrt{|p|}a_2{}' & -\sqrt{|p|}a_1{}' & \sqrt{|p|}a_0{}' \\ -\sqrt{|p|}a_0{}' & \sqrt{|p|}a_1{}' & \sqrt{|p|}a_2{}' & \sqrt{|p|}a_3{}' & a_0 & -a_1 & -a_2 & -a_3 \\ -\sqrt{|p|}a_1{}' & -\sqrt{|p|}a_0{}' & -\sqrt{|p|}a_3{}' & \sqrt{|p|}a_2{}' & a_1 & a_0 & a_3 & -a_2 \\ -\sqrt{|p|}a_2{}' & \sqrt{|p|}a_3{}' & -\sqrt{|p|}a_0{}' & -\sqrt{|p|}a_1{}' & a_2 & -a_3 & a_0 & a_1 \\ -\sqrt{|p|}a_3{}' & -\sqrt{|p|}a_2{}' & \sqrt{|p|}a_1{}' & -\sqrt{|p|}a_0{}' & a_3 & a_2 & -a_1 & a_0 \end{bmatrix}$$

$$\cong \begin{bmatrix} \mathbb{H}^-(\mathbf{Q}) & \sqrt{|p|}\mathbb{H}^-(\mathbf{Q}') \\ -\sqrt{|p|}\mathbb{H}^-(\mathbf{Q}') & \mathbb{H}^-(\mathbf{Q}) \end{bmatrix}$$

Proof. For the elliptic biquaternion $\mathbf{Q}$ the matrix representation is

$$\mathbb{H}^-(\mathbf{Q}) \cong (a_0 + Ta_0')\xi_0 + (a_1 + Ta_0')\xi_1 + (a_2 + Ta_0')\xi_2 + (a_3 + Ta_0')\xi_3 \tag{3.14}$$

First, let us define the following matrix

$$T = \mu \times \Gamma_0 = \begin{bmatrix} 0 & \Gamma_0\sqrt{|p|} \\ -\Gamma_0\sqrt{|p|} & 0 \end{bmatrix}$$

$$T = \begin{bmatrix} 0 & 0 & 0 & 0 & \sqrt{|p|} & 0 & 0 & 0 \\ 0 & 0 & 0 & 0 & 0 & \sqrt{|p|} & 0 & 0 \\ 0 & 0 & 0 & 0 & 0 & 0 & \sqrt{|p|} & 0 \\ 0 & 0 & 0 & 0 & 0 & 0 & 0 & \sqrt{|p|} \\ -\sqrt{|p|} & 0 & 0 & 0 & 0 & 0 & 0 & 0 \\ 0 & -\sqrt{|p|} & 0 & 0 & 0 & 0 & 0 & 0 \\ 0 & 0 & -\sqrt{|p|} & 0 & 0 & 0 & 0 & 0 \\ 0 & 0 & 0 & -\sqrt{|p|} & 0 & 0 & 0 & 0 \end{bmatrix} \tag{3.15}$$

where the quaternionic units provide the multiplication rules

$$\xi_0{}^2 = -\xi_j{}^2 = I_8 = \xi_0,\ \xi_1\xi_2 = -\xi_3,\ \xi_2\xi_3 = -\xi_1,\ \xi_3\xi_1 = -\xi_2\ \ \xi_2\xi_1 = \xi_3\ ,\ \xi_3\xi_2 = \xi_1\ ,\ \xi_1\xi_3 = \xi_2. \tag{3.16}$$

Also, let us define matrices of ξ_j, $j = 0, 1, 2, 3$ and elliptic matrix μ, 2×2 as

$$\xi_j = \sigma_0 \times \Gamma_j = \begin{bmatrix} \Gamma_j & 0 \\ 0 & \Gamma_j \end{bmatrix}$$

and

$$\mu = \begin{bmatrix} 0 & \sqrt{|p|} \\ -\sqrt{|p|} & 0 \end{bmatrix}.$$

Let us find the 8×8 real matrix representation of the equation given in Theorem 3.2. Firstly,

$$a_0\xi_0 = \begin{bmatrix} a_0\Gamma_0 & 0 \\ 0 & a_0\Gamma_0 \end{bmatrix} = \begin{bmatrix} a_0 & 0 & 0 & 0 & 0 & 0 & 0 & 0 \\ 0 & a_0 & 0 & 0 & 0 & 0 & 0 & 0 \\ 0 & 0 & a_0 & 0 & 0 & 0 & 0 & 0 \\ 0 & 0 & 0 & a_0 & 0 & 0 & 0 & 0 \\ 0 & 0 & 0 & 0 & a_0 & 0 & 0 & 0 \\ 0 & 0 & 0 & 0 & 0 & a_0 & 0 & 0 \\ 0 & 0 & 0 & 0 & 0 & 0 & a_0 & 0 \\ 0 & 0 & 0 & 0 & 0 & 0 & 0 & a_0 \end{bmatrix}$$

and

$$(Ta_0{}')\xi_0 = \begin{bmatrix} 0 & {\Gamma_0}^2\sqrt{|p|}a_0{}' \\ -{\Gamma_0}^2\sqrt{|p|}a_0{}' & 0 \end{bmatrix}.$$

So, we can clearly write that

$$(Ta_0{}')\xi_0 = \begin{bmatrix} 0 & 0 & 0 & 0 & a_0{}'\sqrt{|p|} & 0 & 0 & 0 \\ 0 & 0 & 0 & 0 & 0 & a_0{}'\sqrt{|p|} & 0 & 0 \\ 0 & 0 & 0 & 0 & 0 & 0 & a_0{}'\sqrt{|p|} & 0 \\ 0 & 0 & 0 & 0 & 0 & 0 & 0 & a_0{}'\sqrt{|p|} \\ -a_0{}'\sqrt{|p|} & 0 & 0 & 0 & 0 & 0 & 0 & 0 \\ 0 & -a_0{}'\sqrt{|p|} & 0 & 0 & 0 & 0 & 0 & 0 \\ 0 & 0 & -a_0{}'\sqrt{|p|} & 0 & 0 & 0 & 0 & 0 \\ 0 & 0 & 0 & -a_0{}'\sqrt{|p|} & 0 & 0 & 0 & 0 \end{bmatrix}.$$

Thus, we obtain the matrix sum of the two equations as follows:

$$a_0\xi_0 + (Ta_0{}')\xi_0 = \begin{bmatrix} q_0 & 0 & 0 & 0 & \sqrt{|p|}a_0{}' & 0 & 0 & 0 \\ 0 & q_0 & 0 & 0 & 0 & \sqrt{|p|}a_0{}' & 0 & 0 \\ 0 & 0 & q_0 & 0 & 0 & 0 & \sqrt{|p|}a_0{}' & 0 \\ 0 & 0 & 0 & q_0 & 0 & 0 & 0 & \sqrt{|p|}a_0{}' \\ -\sqrt{|p|}a_0{}' & 0 & 0 & 0 & q_0 & 0 & 0 & 0 \\ 0 & -\sqrt{|p|}a_0{}' & 0 & 0 & 0 & q_0 & 0 & 0 \\ 0 & 0 & -\sqrt{|p|}a_0{}' & 0 & 0 & 0 & q_0 & 0 \\ 0 & 0 & 0 & -\sqrt{|p|}a_0{}' & 0 & 0 & 0 & q_0 \end{bmatrix}.$$

Similarly, their matrices of addition can be calculated according to other unit bases of the elliptic biquaternion. Hence, we obtain the matrix representation in 8×8 dimension of the elliptic biquaternion $\mathbf{Q}$ as follows:

$$\mathbb{H}^-(\mathbf{Q}) \cong \begin{bmatrix} a_0 & -a_1 & -a_2 & -a_3 & \sqrt{|p|}a'_0 & -\sqrt{|p|}a'_1 & -\sqrt{|p|}a'_2 & -\sqrt{|p|}a'_3 \\ a_1 & a_0 & a_3 & -a_2 & \sqrt{|p|}a'_1 & \sqrt{|p|}a'_0 & \sqrt{|p|}a'_3 & -\sqrt{|p|}a'_2 \\ a_2 & -a_3 & a_0 & a_1 & \sqrt{|p|}a'_2 & -\sqrt{|p|}a'_3 & \sqrt{|p|}a'_0 & \sqrt{|p|}a'_1 \\ a_3 & a_2 & -a_1 & a_0 & \sqrt{|p|}a'_3 & \sqrt{|p|}a'_2 & -\sqrt{|p|}a'_1 & \sqrt{|p|}a'_0 \\ -\sqrt{|p|}a'_0 & \sqrt{|p|}a'_1 & \sqrt{|p|}a'_2 & \sqrt{|p|}a'_3 & a_0 & -a_1 & -a_2 & -a_3 \\ -\sqrt{|p|}a'_1 & -\sqrt{|p|}a'_0 & -\sqrt{|p|}a'_3 & \sqrt{|p|}a'_2 & a_1 & a_0 & a_3 & -a_2 \\ -\sqrt{|p|}a'_2 & \sqrt{|p|}a'_3 & -\sqrt{|p|}a'_0 & -\sqrt{|p|}a'_1 & a_2 & -a_3 & a_0 & a_1 \\ -\sqrt{|p|}a'_3 & -\sqrt{|p|}a'_2 & \sqrt{|p|}a'_1 & -\sqrt{|p|}a'_0 & a_3 & a_2 & -a_1 & a_0 \end{bmatrix}$$

and thus, we get

$$\mathbb{H}^-(\mathbf{Q}) \cong \begin{bmatrix} \mathbb{H}^-(\mathbf{Q}) & \sqrt{|p|}\mathbb{H}^-(\mathbf{Q}') \\ -\sqrt{|p|}\mathbb{H}^-(\mathbf{Q}') & \mathbb{H}^-(\mathbf{Q}) \end{bmatrix}.$$

Q.E.D.

4 Lorentz transformations with elliptic biquaternions

Lorentz transformation in physics is named after Dutch physicist H. Lorentz. The transformations explain how the speed of light independent of the frame of reference is to be observed the measurements of space and time measured by the two observers being related. Lorentz transformations are linear transformations in concordance with special relativity. Every elliptic biquaternion $\mathbf{R}$ that combines the cartesian position vector $\vec{r} = r_1\hat{i} + r_2\hat{j} + r_3\hat{k}$ with the time t can be written as follows:

$$\mathbf{R} = ct + I\boldsymbol{r} = cte_0 + I(r_1e_1 + r_2e_2 + r_3e_3) = R_0e_0 + R_1e_1 + R_2e_2 + R_3e_3 \tag{4.1}$$

where c is the speed of light in four-dimensional space, $\boldsymbol{r}$ is the pure real quaternion. Considering the left Hamiltonian matrix and Theorem 3.2 for the elliptic biquaternion $\mathbf{R} = ct + I\boldsymbol{r}$ the following theorem can be given.

Theorem 4.1. Let us consider an elliptical biquaternion $\mathbf{R} = ct + I\mathbf{r}$ that combines vector of position and time in four-dimensional space. The elliptic biquaternionic representation of $\mathbb{H}^-(\mathbf{R})$ is as follows

$$\mathbb{H}^-(\mathbf{R}) \cong \begin{bmatrix} ct\Gamma_0 & \sqrt{|p|}\mathbb{H}^-(\mathbf{R}) \\ -\sqrt{|p|}\mathbb{H}^-(\mathbf{R}) & ct\Gamma_0 \end{bmatrix}$$

where c indicate the speed of light.

Proof. Matrices corresponding to the base elements of the pure elliptic biquaternion $\boldsymbol{R}$ are written as follows

$$\Gamma_1 = \begin{bmatrix} \sigma_2 & 0 \\ 0 & -\sigma_2 \end{bmatrix}, \quad \Gamma_2 = \begin{bmatrix} 0 & -\sigma_0 \\ \sigma_0 & 0 \end{bmatrix}, \quad \Gamma_3 = \begin{bmatrix} 0 & \sigma_2 \\ \sigma_2 & 0 \end{bmatrix}.$$

Considering these matrices the equality, we get

$$\mathbb{H}^-(\mathbf{R}) = r_1\Gamma_1 + r_2\Gamma_2 + r_3\Gamma_3$$

$$\mathbb{H}^-(\mathbf{R}) = \begin{bmatrix} 0 & -r_1 & 0 & 0 \\ r_1 & 0 & 0 & 0 \\ 0 & 0 & 0 & r_1 \\ 0 & 0 & -r_1 & 0 \end{bmatrix} + \begin{bmatrix} 0 & 0 & -r_2 & 0 \\ 0 & 0 & 0 & -r_2 \\ r_2 & 0 & 0 & 0 \\ 0 & r_2 & 0 & 0 \end{bmatrix} + \begin{bmatrix} 0 & 0 & 0 & -r_3 \\ 0 & 0 & r_3 & 0 \\ 0 & -r_3 & 0 & 0 \\ r_3 & 0 & 0 & 0 \end{bmatrix}$$

$$= \begin{bmatrix} 0 & -r_1 & -r_2 & -r_3 \\ r_1 & 0 & r_3 & -r_2 \\ r_2 & -r_3 & 0 & r_1 \\ r_3 & r_2 & -r_1 & 0 \end{bmatrix}.$$

Also, from the equalities 3.15 and 3.16, we obtain the real matrix in dimension 8×8 as follows

$$\mathbb{H}^{-}(\mathbf{R}) = ct\xi_0 + (Tr_1)\xi_1 + (Tr_2)\xi_2 + (Tr_3)\xi_3 \tag{4.2}$$

$$\mathbb{H}^{-}(\mathbf{R}) \cong \begin{bmatrix} ct & 0 & 0 & 0 & 0 & -r_1\sqrt{|p|} & -r_2\sqrt{|p|} & -r_3\sqrt{|p|} \\ 0 & ct & 0 & 0 & r_1\sqrt{|p|} & 0 & r_3\sqrt{|p|} & -r_2\sqrt{|p|} \\ 0 & 0 & ct & 0 & r_2\sqrt{|p|} & -r_3\sqrt{|p|} & 0 & r_1\sqrt{|p|} \\ 0 & 0 & 0 & ct & r_3\sqrt{|p|} & r_2\sqrt{|p|} & -r_1\sqrt{|p|} & 0 \\ 0 & r_1\sqrt{|p|} & r_2\sqrt{|p|} & r_3\sqrt{|p|} & ct & 0 & 0 & r_3\sqrt{|p|} \\ -r_1\sqrt{|p|} & 0 & -r_3\sqrt{|p|} & r_2\sqrt{|p|} & 0 & ct & -r_3\sqrt{|p|} & 0 \\ -r_2\sqrt{|p|} & r_3\sqrt{|p|} & 0 & -r_1\sqrt{|p|} & 0 & r_3\sqrt{|p|} & ct & 0 \\ -r_3\sqrt{|p|} & -r_2\sqrt{|p|} & r_1\sqrt{|p|} & 0 & -r_3\sqrt{|p|} & 0 & 0 & ct \end{bmatrix}.$$

Hence, 2×2 elliptical matrix representation of the resulting matrix is as follows:

$$\mathbb{H}^{-}(\mathbf{R}) \cong \begin{bmatrix} ct\Gamma_0 & \sqrt{|p|}\mathbb{H}^{-}(\mathbf{R}) \\ -\sqrt{|p|}\mathbb{H}^{-}(\mathbf{R}) & ct\Gamma_0 \end{bmatrix}.$$

Q.E.D.

As a result of the relativistic transformation relation of the elliptic biquaternion $\mathbf{R} = ct + I\boldsymbol{r}$ Lorentz transformation equations can be obtained.
Firstly, let us the unit elliptic biquaternion $\mathbf{Q}$ which can be define as

$$\mathbf{Q} = a_0 + I\boldsymbol{a}' = A_0e_0 + A_1e_1 + A_2e_2 + A_3e_3 = A_0 + \boldsymbol{A}. \tag{4.3}$$

Considering Theorem 2.3 we state the unit elliptic biquaternion as follows:

$$\mathbf{Q} = \cosh(p\frac{\theta_p}{2}) + \frac{1}{I}\hat{q}\sinh(p\frac{\theta_p}{2}) = \cosh(p\frac{\theta_p}{2}) + \frac{1}{I}I\frac{\vec{v}}{\sqrt{<\vec{v},\vec{v}>}}\sinh(p\frac{\theta_p}{2}). \tag{4.4}$$

Here, we consider a speed $\vec{v} = v_1e_1 + v_2e_2 + v_3e_3$ which is in the opposite direction. Thus, from 4.3 and 4.4 the following equation holds:

$$a_0 = A_0 = \cosh(p\frac{\theta_p}{2})\ ,\ \boldsymbol{a}' = \frac{1}{I}\frac{\vec{v}}{\nu}\sinh(p\frac{\theta_p}{2}) = \frac{\hat{q}}{p}\sinh(p\frac{\theta_p}{2}),\ \boldsymbol{A} = \frac{\vec{v}}{\nu}\sinh(p\frac{\theta_p}{2}). \tag{4.5}$$

On the other hand from the known Lorentz coordinate transformations, we get

$$\begin{aligned} \cosh(p\theta_p) &= \frac{|c|}{\sqrt{c^2-\nu^2}} = \frac{|c|}{|c|\sqrt{1-\frac{\nu^2}{c^2}}} = \frac{1}{1-|p|\,\nu^2} = \beta \\ \sinh(p\theta_p) &= \frac{|\nu|}{\sqrt{c^2-\nu^2}} = \frac{|\nu|}{|c|\sqrt{1-|p|\,v^2}} = \mp\frac{\nu}{c\sqrt{1-p\nu^2}} = \mp\frac{\beta}{c}\nu \end{aligned} \tag{4.6}$$

where $|p| = \frac{1}{c^2}$ and $\sqrt{<\vec{v},\vec{v}>} = \nu$. Also, from the 4.1 and we obtain

$$\begin{aligned} a_0 = A_0,\quad A = Ia',\quad A_1 = Ia_1',\quad A_2 = Ia_2', \\ A_3 = Ia_3',\quad a_0'e_0 + \quad a_1'e_1 + a_2'e_2 + a_3'e_3 = a', R_0 = ct,\quad \boldsymbol{R} = I\boldsymbol{r}. \end{aligned} \tag{4.7}$$

The elliptic biquaternion obtained from 4.1 as a result of the relation of the relativistic transformation of elliptic biquaternion $\mathbf{R} = ct + I\boldsymbol{r}$ can be written as $\mathbf{R}' = ct + I\boldsymbol{r'}$. The relativistic transformation relation of elliptic biquaternion $\mathbf{R}$ is expressed as follows

$$\mathbf{R}' = \mathbf{Q}\mathbf{R}\bar{\mathbf{Q}}^{*}. \tag{4.8}$$

This equality is rewritten using equations 3.15 and Theorem 4.1 as

$$\begin{aligned}\mathbb{H}^{-}(\mathbf{R}') &= \mathbb{H}^{-}(\mathbf{Q})\mathbb{H}^{-}(\mathbf{R})\mathbb{H}^{-}(\bar{\mathbf{Q}})^{*} \\ &= (a_0\xi_0 + (Ta_1)\xi_1 + (Ta_2)\xi_2 + (Ta_3)\xi_3)(t\xi_0 + (Tr_1)\xi_1 + (Tr_2)\xi_2 \\ &\quad +(Tr_3)\xi_3)(a_0\xi_0 + (Ta_1)\xi_1 + (Ta_2)\xi_2 + (Ta_3)\xi_3)\end{aligned} \tag{4.9}$$

or in the other way

$$\mathbb{H}^{-}(\mathbf{R}') \cong \begin{bmatrix} a_0\Gamma_0 & -\sqrt{|p|}\mathbb{H}^{-}(\mathbf{Q}) \\ -\sqrt{|p|}\mathbb{H}^{-}(\mathbf{Q}) & a_0\Gamma_0 \end{bmatrix}\begin{bmatrix} ct\Gamma_0 & \sqrt{|p|}\mathbb{H}^{-}(\boldsymbol{R}) \\ -\sqrt{|p|}\mathbb{H}^{-}(\boldsymbol{R}) & ct\Gamma_0 \end{bmatrix}\begin{bmatrix} a_0\Gamma_0 & \sqrt{|p|}\mathbb{H}^{-}(\mathbf{Q}) \\ -\sqrt{|p|}\mathbb{H}^{-}(\mathbf{Q}) & a_0\Gamma_0 \end{bmatrix} \tag{4.10}$$

is represented. Also by equality 4.8 we write

$$\begin{aligned}\mathbf{R}' &= (A_0 + \boldsymbol{A})(A_0R_0 + R_0\boldsymbol{A} + A_0\boldsymbol{R} - \langle \boldsymbol{R}, \boldsymbol{A}\rangle + (\boldsymbol{R}\wedge A) \\ &= {A_0}^2R_0 + A_0R_0\boldsymbol{A} + {A_0}^2\boldsymbol{R} - A_0 < \boldsymbol{R}, \boldsymbol{A} > +A_0 < \boldsymbol{R}, \boldsymbol{A} > +A_0R_0\boldsymbol{A} \\ &\quad - R_0 < \boldsymbol{A}, \boldsymbol{A} > -A_0 < \boldsymbol{A}, \boldsymbol{R} > \\ &= A_0\langle \boldsymbol{A}, \boldsymbol{R}\rangle - \boldsymbol{A}\langle \boldsymbol{R}, \boldsymbol{A}\rangle - \langle \boldsymbol{A}, \boldsymbol{R}\wedge \boldsymbol{A}\rangle + \boldsymbol{A}\wedge(\boldsymbol{R}\wedge\boldsymbol{A}).\end{aligned}$$

Since there is $\langle \boldsymbol{A}, \boldsymbol{R}\wedge\boldsymbol{A}\rangle = 0$ we reach the following general equality for $\mathbf{R}'$:

$$\begin{aligned}\mathbf{R}' = {A_0}^2R_0 + 2A_0R_0A + {A_0}^2\boldsymbol{R} - A_0 < \boldsymbol{R}, \boldsymbol{A} > -R_0 < \boldsymbol{A}, \boldsymbol{A} > \\ -A_0 < \boldsymbol{A}, \boldsymbol{R} > -\boldsymbol{A} < \boldsymbol{R}, \boldsymbol{A} > +\boldsymbol{A}\wedge(\boldsymbol{R}\wedge\boldsymbol{A}).\end{aligned} \tag{4.11}$$

From the properties of the vector product, we get

$$\boldsymbol{A}\wedge(\boldsymbol{R}\wedge\boldsymbol{A}) = \boldsymbol{R} < \boldsymbol{A}, \boldsymbol{A} > -\boldsymbol{A} < \boldsymbol{A}, \boldsymbol{R} >= \boldsymbol{A}^2\boldsymbol{R} - \boldsymbol{A} < \boldsymbol{A}, \boldsymbol{R} > .$$

If this equality is written in the equation 4.11, we obtain

$$\mathbf{R}' = {A_0}^2R_0 + 2A_0R_0\boldsymbol{A} + {A_0}^2\boldsymbol{R} - 2A_0 < \boldsymbol{A}, \boldsymbol{R} > -R_0\boldsymbol{A}^2 - 2\boldsymbol{A} < \boldsymbol{A}, \boldsymbol{R} > +\boldsymbol{A}^2\boldsymbol{R}.$$

By substituting 4.7 in this equation, we reach the following equation

$$\begin{aligned}\mathbf{R}' &= {A_0}^2ct - ct(I\boldsymbol{a'})^2 - 2A_0I^2 < \boldsymbol{r}, \boldsymbol{a'} > +{A_0}^2(I\boldsymbol{r}) + 2A_0ct(I\boldsymbol{a'}) \\ &\quad - 2I^3\boldsymbol{a'} < \boldsymbol{a'}, \boldsymbol{r} > +(I\boldsymbol{a'})^2(I\boldsymbol{r}) \\ &= ({A_0}^2 - p(\boldsymbol{a'})^2)ct - 2A_0p < \boldsymbol{r}, \boldsymbol{a'} > +I\boldsymbol{r}({A_0}^2 + p(\boldsymbol{a'})^2) + 2IctA_0\boldsymbol{a'} \\ &\quad - 2Ip\boldsymbol{a'} < \boldsymbol{a'}, \boldsymbol{r} >.\end{aligned}$$

In this expression, the terms containing and not containing I ($I^2 = p < 0$) must be matched so that both sides of equality can be equal to each other. Thus, for relativistic transformation relation of space and time as the equation $\mathbf{R}' = ct' + I\boldsymbol{r}'$ can be written we obtain the following equations

$$ct' = ({A_0}^2 - p(\boldsymbol{a'})^2)ct - 2A_0 p < \boldsymbol{r}, \boldsymbol{a'} >$$
$$t' = ({A_0}^2 - p(\boldsymbol{a'})^2)t - 2A_0 \frac{p}{c} < \boldsymbol{r}, \boldsymbol{a'} > \tag{4.12}$$

and

$$I\boldsymbol{r'} = I\boldsymbol{r}({A_0}^2 + p(\boldsymbol{a'})^2) + 2IctA_0\boldsymbol{a'} - 2Ip\boldsymbol{a'} < \boldsymbol{a'}, \boldsymbol{r} >$$
$$\boldsymbol{r'} = \boldsymbol{r}({A_0}^2 + p(\boldsymbol{a'})^2) + 2ctA_0\boldsymbol{a'} - 2p\boldsymbol{a'} < \boldsymbol{a'}, \boldsymbol{r} > . \tag{4.13}$$

For the elliptic biquaternion $\mathbf{Q}$ the definitions given in the equations 4.3-4.4 using in the equation 4.12 we obtain

$$t' = (\cosh^2(p\frac{\theta_p}{2}) + \sinh^2(p\frac{\theta_p}{2})t - 2\cosh(p\frac{\theta_p}{2})\frac{p}{2}\frac{1}{c\,p} < \boldsymbol{r}, \hat{q} > \sinh(p\frac{\theta_p}{2}). \tag{4.14}$$

Furthermore, using the equations of 4.6 and the known hyperbolic trigonometric function we obtain the expression 4.14 as follows:

$$t' = t\cosh(p\theta_p) - \frac{I}{c}\sinh(p\theta_p)\frac{<\boldsymbol{r}, \vec{v}>}{\sqrt{<\vec{v}, \vec{v}>}}.$$

Here

$$\hat{q} = I\frac{\vec{v}}{\sqrt{<\vec{v}, \vec{v}>}} \tag{4.15}$$

therefore we obtain

$$t' = t\beta - I\frac{\beta}{c}\frac{|\nu|}{c}\frac{<\boldsymbol{r}, \vec{v}>}{\sqrt{<\vec{v}, \vec{v}>}} = \beta(t - \frac{I}{c^2} < \boldsymbol{r}, \vec{v} >). \tag{4.16}$$

Similarly, for the equality $\boldsymbol{r'}$ given in 4.13 we get

$$\begin{aligned}\boldsymbol{r'} &= \boldsymbol{r}(\cosh^2(p\frac{\theta_p}{2}) + p < \hat{q}, \hat{q} > \sinh^2(p\frac{\theta_p}{2}) + 2ct\cosh(p\frac{\theta_p}{2})\frac{\hat{q}}{p}\sinh(p\frac{\theta_p}{2}) \\ &-2p\frac{\hat{q}}{p}\sinh(p\frac{\theta_p}{2}) < \frac{\hat{q}}{p}\sinh(p\frac{\theta_p}{2}), \boldsymbol{r} > .\end{aligned} \tag{4.17}$$

Writing the equation 4.17 in the last equation we find

$$\begin{aligned}\boldsymbol{r'} &= \boldsymbol{r} + ct\frac{I}{p}\frac{\vec{v}}{\sqrt{<\vec{v}, \vec{v}>}}2\cosh(p\frac{\theta_p}{2})\sinh(p\frac{\theta_p}{2}) - 2\frac{1}{p}\sinh^2(p\frac{\theta_p}{2})I^2\frac{<\vec{v}, \boldsymbol{r}>}{\sqrt{<\vec{v}, \vec{v}>}\sqrt{<\vec{v}, \vec{v}>}} \\ &-2\sinh^2(p\frac{\theta_p}{2})\frac{I^2}{p}\frac{1}{\sqrt{<v,v>}\sqrt{<v,v>}}\vec{v} < \vec{v}, \boldsymbol{r} >\end{aligned}$$

with the help of trigonometric identities we obtain

$$\boldsymbol{r'} = \boldsymbol{r} + \frac{1}{I}\frac{1}{\sqrt{1-p\nu^2}}\vec{v}t - \frac{1}{\sqrt{1-p\nu^2}}\frac{\vec{v}<\vec{v}, \boldsymbol{r}>}{\nu^2} + \frac{\vec{v}<\vec{v}, \boldsymbol{r}>}{\nu^2}.$$

Thus

$$\boldsymbol{r'} = x + \frac{1}{I}\frac{1}{\sqrt{1-|p|v^2}}\vec{v}t \pm \frac{x}{\sqrt{1-|p|v^2}} \mp x = \frac{1}{I}\frac{1}{\sqrt{1-|p|v^2}}\vec{v}t + \frac{x}{\sqrt{1-|p|v^2}}$$

is obtained. Consequently, the equation is obtained as follows:

$$\boldsymbol{r}' = \frac{I}{I^2}\frac{\vec{v}t}{\sqrt{1-|p|\,v^2}} + \frac{x}{\sqrt{1-|p|\,v^2}} = \frac{1}{\sqrt{1-|p|\,v^2}}(x - \frac{I}{|p|}\vec{v}t) = \beta(x - \frac{I}{|p|}\vec{v}t). \tag{4.18}$$

Here is $\boldsymbol{r} = +x$, $v = -x$, $< v, v >= \mp\nu^2$. These obtained equations for t' and $\boldsymbol{r'}$ are in conformity with the usual Lorentz transformation equations. It is also possible to give the matrix equality 4.9 which defines relativistic transformation carrying out with 4-dimensional matrix representations. For this purpose, if the elliptic biquaternion matrix definition in equality 3.4 is used, the equality in the form $\mathbf{R}' = \mathbf{Q}\mathbf{R}\bar{\mathbf{Q}}^*$ can be written. We give the matrix representation of this transformation using the matrix multiplication relations 3.5 and 3.9 since the matrices do not provide the commutative property. We get a unit elliptic biquaternion as $\mathbf{Q} = a_0 + I\boldsymbol{a}'$ $(a_1 = a_2 = a_3 = a_0' = 0)$ and $I^2 = p < 0$ then

$$\mathbb{H}^-(\mathbf{R}') = \mathbb{H}^-(\mathbf{Q})\ \mathbf{R}\mathbb{H}^-(\bar{\mathbf{Q}}^*) = \mathbb{H}^-(\mathbf{Q})\mathbb{H}^+(\bar{\mathbf{Q}}^*)\ \mathbf{R} \tag{4.19}$$

$$\begin{bmatrix} R'_0 \\ \boldsymbol{R'} \end{bmatrix} = \begin{bmatrix} A_0 & -\boldsymbol{A} \\ \boldsymbol{A^T} & A_0 I_3 + \tilde{\boldsymbol{A}} \end{bmatrix} \begin{bmatrix} R_0 & 0 \\ 0 & R_0 I_3 \end{bmatrix} \begin{bmatrix} A_0 \\ \boldsymbol{A} \end{bmatrix}$$

is obtained. This equation can be written more clearly in type 4×4 as follows:

$$\begin{bmatrix} R'_0 \\ R'_1 \\ R'_2 \\ R'_3 \end{bmatrix} = \begin{bmatrix} A_0 & -A_1 & -A_2 & -A_3 \\ A_1 & A_0 & A_3 & -A_2 \\ A_2 & -A_3 & A_0 & A_1 \\ A_3 & A_2 & -A_1 & A_0 \end{bmatrix} \begin{bmatrix} A_0 & -A_1 & -A_2 & -A_3 \\ A_1 & A_0 & -A_3 & A_2 \\ A_2 & A_3 & A_0 & -A_1 \\ A_3 & -A_2 & A_1 & A_0 \end{bmatrix} \begin{bmatrix} R_0 \\ R_1 \\ R_2 \\ R_3 \end{bmatrix} \tag{4.20}$$

the equation

$$\mathbf{U} = \begin{bmatrix} {A_0}^2 - {A_1}^2 - {A_2}^2 - {A_3}^2 & -2A_0A_1 & -2A_0A_2 & -2A_0A_3 \\ 2A_0A_1 & {A_0}^2 - {A_1}^2 + {A_2}^2 + {A_3}^2 & -2A_1A_2 & -2A_1A_3 \\ 2A_0A_2 & -2A_1A_2 & {A_0}^2 + {A_1}^2 - {A_2}^2 + {A_3}^2 & -2A_2A_3 \\ 2A_0A_3 & -2A_1A_3 & -2A_2A_3 & {A_0}^2 + {A_1}^2 + {A_2}^2 - {A_3}^2 \end{bmatrix}$$

is concluded. Since the elliptic biquaternion $\mathbf{Q}$ is the unit elliptic biquaternion the above matrix relation can be written in a simpler form as

$$\begin{bmatrix} R'_0 \\ R'_1 \\ R'_2 \\ R'_3 \end{bmatrix} = \begin{bmatrix} 2{A_0}^2 - 1 & -2A_0A_1 & -2A_0A_2 & -2A_0A_3 \\ 2A_0A_1 & 1 - 2{A_1}^2 & -2A_1A_2 & -2A_1A_3 \\ 2A_0A_2 & -2A_1A_2 & 1 - 2{A_2}^2 & -2A_2A_3 \\ 2A_0A_3 & -2A_1A_3 & -2A_2A_3 & 1 - 2{A_3}^2 \end{bmatrix} \begin{bmatrix} R_0 \\ R_1 \\ R_2 \\ R_3 \end{bmatrix}$$

$$\begin{bmatrix} R'_0 \\ Ir'_1 \\ Ir'_2 \\ Ir'_3 \end{bmatrix} = \begin{bmatrix} 2{A_0}^2 - 1 & -2A_0A_1 & -2A_0A_2 & -2A_0A_3 \\ 2A_0A_1 & 1 - 2{A_1}^2 & -2A_1A_2 & -2A_1A_3 \\ 2A_0A_2 & -2A_1A_2 & 1 - 2{A_2}^2 & -2A_2A_3 \\ 2A_0A_3 & -2A_1A_3 & -2A_2A_3 & 1 - 2{A_3}^2 \end{bmatrix} \begin{bmatrix} R_0 \\ Ir_1 \\ Ir_2 \\ Ir_3 \end{bmatrix}.$$

In this equation we get

$$\mathbf{U}=\begin{bmatrix} 2{A_0}^2-1 & -2A_0A_1 & -2A_0A_2 & -2A_0A_3 \\ 2A_0A_1 & 1-2{A_1}^2 & -2A_1A_2 & -2A_1A_3 \\ 2A_0A_2 & -2A_1A_2 & 1-2{A_2}^2 & -2A_2A_3 \\ 2A_0A_3 & -2A_1A_3 & -2A_2A_3 & 1-2{A_3}^2 \end{bmatrix}.$$

If the matrix equality is defined as above then using definitions 4.6 is obtained as

$$\mathbf{U}=\begin{bmatrix} \beta & \pm\frac{\beta}{c}v_1 & \pm\frac{\beta}{c}v_2 & \pm\frac{\beta}{c}v_3 \\ \mp\frac{\beta}{c}v_1 & 1-(\beta-1)\frac{{v_1}^2}{\mp\nu^2} & -(\beta-1)\frac{v_1v_2}{\mp\nu^2} & -(\beta-1)\frac{v_1v_3}{\mp\nu^2} \\ \mp\frac{\beta}{c}v_2 & -(\beta-1)\frac{v_1v_2}{\pm\nu^2} & 1-(\beta-1)\frac{{v_2}^2}{\mp\nu^2} & -(\beta-1)\frac{v_2v_3}{\mp\nu^2} \\ \mp\frac{\beta}{c}v_3 & -(\beta-1)\frac{v_1v_3}{\pm\nu^2} & -(\beta-1)\frac{v_2v_3}{\mp\nu^2} & 1-(\beta-1)\frac{{v_3}^2}{\mp\nu^2} \end{bmatrix}.$$

Thus, it is possible to represent equality 4.19 as

$$\mathbb{H}^-(\mathbf{R}')=\mathbf{U}\ \mathbf{R}.$$

In addition, it is possible to indicate the equation defined in 4.19 with eight-dimensional real matrices.

We can write as $\mathbb{H}^-(\mathbf{R}')=\mathbb{H}^-(\mathbf{Q})\ \mathbf{R}\ \mathbb{H}^-(\bar{\mathbf{Q}}^*)=\mathbb{H}^-(\mathbf{Q})\mathbb{H}^-(\overset{*}{\breve{\bar{\mathbf{Q}}}})\ \mathbf{R}$. This equation can be written as

$$\mathbb{H}^-(\mathbf{R}')=\begin{bmatrix} a_0 & 0 & a'_0 & -a' \\ 0 & a_0I_3 & a'T & a'_0I_3-\tilde{a}' \\ -a'_0 & a' & a_0 & 0 \\ -a'^T & -a'_0I_3-\tilde{a}' & 0 & a_0I_3 \end{bmatrix}\begin{bmatrix} a_0 & 0 & a'_0 & -a' \\ 0 & a_0I_3 & a'^T & {a'_0}^TI_3-\tilde{a}' \\ -a'_0 & a' & a_0 & 0 \\ -a'^T & -a'_0I_3+\tilde{a}' & 0 & a_0I_3 \end{bmatrix}\begin{bmatrix} R_0 \\ 0 \\ 0 \\ R \end{bmatrix}.$$

Also, it can be obtained more clearly as

$$\begin{bmatrix} ct' \\ 0 \\ 0 \\ 0 \\ 0 \\ r'_1 \\ r'_2 \\ r'_3 \end{bmatrix}=\begin{bmatrix} a_0 & 0 & 0 & 0 & a'_0 & -a'_1 & -a'_2 & -a'_3 \\ 0 & a_0 & 0 & 0 & a'_1 & a'_0 & a'_3 & -a'_2 \\ 0 & 0 & a_0 & 0 & a'_2 & -a'_3 & a'_0 & a'_1 \\ 0 & 0 & 0 & a_0 & a'_3 & a'_2 & -a'_1 & a'_0 \\ -a'_0 & a'_1 & a'_2 & a'_3 & a_0 & 0 & 0 & 0 \\ -a'_1 & a'_0 & -a'_3 & a'_2 & 0 & a_0 & 0 & 0 \\ -a'_2 & a'_3 & a'_0 & -a'_1 & 0 & 0 & a_0 & 0 \\ -a'_3 & -a'_2 & a'_1 & -a'_0 & 0 & 0 & 0 & a_0 \end{bmatrix}\begin{bmatrix} a_0 & 0 & 0 & 0 & a'_0 & -a'_1 & -a'_2 & -a'_3 \\ 0 & a_0 & 0 & 0 & a'_1 & a'_0 & -a'_3 & a'_2 \\ 0 & 0 & a_0 & 0 & a'_2 & a'_{33} & a'_0 & -a'_1 \\ 0 & 0 & 0 & a_0 & a'_3 & -a'_2 & a'_1 & a'_0 \\ -a'_0 & a'_1 & a'_2 & a'_3 & a_0 & 0 & 0 & 0 \\ -a'_1 & -a'_0 & a'_3 & -a'_2 & 0 & a_0 & 0 & 0 \\ -a'_2 & -a'_3 & -a'_0 & a'_1 & 0 & 0 & a_0 & 0 \\ -a'_3 & a'_2 & -a'_1 & -a'_0 & 0 & 0 & 0 & a_0 \end{bmatrix}\begin{bmatrix} ct \\ 0 \\ 0 \\ 0 \\ 0 \\ r_1 \\ r_2 \\ r_3 \end{bmatrix}.$$

From here two matrix multiplications

$$\mathbb{H}^-(\mathbf{Q})\,\mathbb{H}^-\left(\breve{\mathbf{Q}}\right)\cong$$

$$\begin{bmatrix} {a_0}^2+{a'_1}^2+{a'_2}^2+{a'_3}^2 & 0 & 0 & 0 & 0 & -2a_0a'_1 & -2a_0a'_2 & -2a_0a'_3 \\ 0 & {a_0}^2+{a'_1}^2-{a'_2}^2-{a'_3}^2 & 2a'_1a'_2 & 2a'_1a'_3 & 2a_0a'_1 & 0 & 0 & 0 \\ 0 & 2a'_1a'_2 & {a_0}^2-{a'_1}^2+{a'_2}^2-{a'_3}^2 & 2a'_2a'_3 & 2a_0a'_2 & 0 & 0 & 0 \\ 0 & 2a'_1a'_3 & 2a'_2a'_3 & {a_0}^2-{a'_1}^2-{a'_2}^2+{a'_3}^2 & 2a_0a'_3 & 0 & 0 & 0 \\ 0 & 2a_0a'_1 & 2a_0a'_2 & 2a_0a'_3 & {a_0}^2+{a'_1}^2+{a'_2}^2+{a'_3}^2 & 0 & 0 & 0 \\ -2a'_1a_0 & 0 & 0 & 0 & 0 & {a_0}^2+{a'_1}^2-{a'_2}^2-{a'_3}^2 & 2a'_1a'_2 & 2a'_1a'_3 \\ -2a'_2a_0 & 0 & 0 & 0 & 0 & 2a'_1a'_2 & {a_0}^2-{a'_1}^2+{a'_2}^2-{a'_3}^2 & 2a'_2a'_3 \\ -2a'_3a_0 & 0 & 0 & 0 & 0 & 2a'_1a'_3 & 2a'_2a'_3 & {a_0}^2-{a'_1}^2-{a'_2}^2+{a'_3}^2 \end{bmatrix}$$

is obtained. We can give this equation more clearly as follows:

$$\mathbb{H}^-(\mathbf{Q})\,\mathbb{H}^-\left(\breve{\mathbf{Q}}\right)\cong$$

$$\begin{bmatrix} \frac{(p-1)a_0{}^2+1}{p} & 0 & 0 & 0 & 0 & -2a_0a'_1 & -2a_0a'_2 & -2a_0a'_3 \\ 0 & \frac{a_0{}^2(p+1)+2pa'^2{}_1-1}{p} & 2a'_1a'_2 & 2a'_1a'_3 & 2a_0a'_1 & 0 & 0 & 0 \\ 0 & 2a'_1a'_2 & \frac{a_0{}^2(p+1)+2pa'_2{}^2-1}{p} & 2a'_2a'_3 & 2a_0a'_2 & 0 & 0 & 0 \\ 0 & 2a'_1a'_3 & 2a'_2a'_3 & \frac{a_0{}^2(p+1)+2pa'_3{}^2-1}{p} & 2a_0a'_3 & 0 & 0 & 0 \\ 0 & 2a_0a'_1 & 2a_0a'_2 & 2a'_3a_0 & \frac{(p-1)a_0{}^2+1}{p} & 0 & 0 & 0 \\ -2a_0a'_1 & 0 & 0 & 0 & 0 & \frac{a_0{}^2(p+1)+2pa'^2{}_1-1}{p} & 2a'_1a'_2 & 2a'_1a'_3 \\ -2a_0a'_2 & 0 & 0 & 0 & 0 & 2a'_1a'_2 & \frac{a_0{}^2(p+1)+2pa'_2{}^2-1}{p} & 2a'_2a'_3 \\ -2a_0a'_3 & 0 & 0 & 0 & 0 & 2a'_1a'_3 & 2a'_2a'_3 & \frac{a_0{}^2(p+1)+2pa'_3{}^2-1}{p} \end{bmatrix} . \tag{4.21}$$

For the real components of the elliptic biquaternion as can be seen from 4.6;

$$a_0 = \cosh(\frac{I\theta_p}{2}) \ , \ a_1{}' = \frac{1}{I}\frac{\vec{v}_1}{\nu}\sinh(\frac{I\theta_p}{2}) \ \ , \ a_2{}' = \frac{1}{I}\frac{\vec{v}_2}{\nu}\sinh(\frac{I\theta_p}{2}) \ , \ a'_3 = \frac{1}{I}\frac{\vec{v}_3}{\nu}\sinh(\frac{I\theta_p}{2})$$

and the transformation matrix $\mathbf{U}$ that provides definitions given in 4.6 is

$$\mathbf{U} \cong \begin{bmatrix} \frac{(\beta+1)(p-1)+2}{2p} & 0 & 0 & 0 & 0 & \pm\frac{1}{I}\frac{\beta}{c}v_1 & \pm\frac{1}{I}\frac{\beta}{c}v_2 & \pm\frac{1}{I}\frac{\beta}{c}v_3 \\ 0 & \frac{(\beta+1)(p+1)+2(\beta-1)\frac{v_1{}^2}{\mp\nu^2}-2}{2p} & \frac{(\beta-1)v_1v_2}{\mp p\nu^2} & \frac{-(\beta-1)v_1v_3}{\mp p\nu^2} & \mp\frac{1}{I}\frac{\beta}{c}v_1 & 0 & 0 & 0 \\ 0 & \frac{(\beta-1)v_1v_2}{\mp pv^2} & \frac{(\beta+1)(p+1)+2(\beta-1)\frac{v_2{}^2}{\mp\nu^2}-2}{2p} & \frac{(\beta-1)v_2v_3}{\mp p\nu^2} & \mp\frac{1}{I}\frac{\beta}{c}v_2 & 0 & 0 & 0 \\ 0 & \frac{(\beta-1)v_1v_3}{\mp pv^2} & \frac{(\beta-1)+v_2v_3}{\mp pv^2} & \frac{(\beta+1)(p+1)+2(\beta-1)\frac{v_3{}^2}{\mp\nu^2}-2}{2p} & \mp\frac{1}{I}\frac{\beta}{c}v_3 & 0 & 0 & 0 \\ 0 & \mp\frac{1}{I}\frac{\beta}{c}v_1 & \mp\frac{1}{I}\frac{\beta}{c}v_2 & \mp\frac{1}{I}\frac{\beta}{c}v_3 & \frac{(p-1)(\beta+1)+2}{2p} & 0 & 0 & 0 \\ \pm\frac{1}{I}\frac{\beta}{c}v_1 & 0 & 0 & 0 & 0 & \frac{(\beta+1)(p+1)+2(\beta-1)\frac{v_1{}^2}{\mp\nu^2}-2}{2p} & \frac{(\beta-1)v_1v_2}{\mp pv^2} & \frac{(\beta-1)v_1v_3}{\mp pv^2} \\ \pm\frac{1}{I}\frac{\beta}{c}v_2 & 0 & 0 & 0 & 0 & \frac{(\beta-1)v_1v_2}{\mp p\nu^2} & \frac{(\beta+1)(p+1)+2(\beta-1)\frac{v_2{}^2}{\nu^2}-2}{2p} & \frac{(\beta-1)v_2v_3}{\mp pv^2} \\ \pm\frac{1}{I}\frac{\beta}{c}v_3 & 0 & 0 & 0 & 0 & \frac{(\beta-1)v_1v_3}{\mp pv^2} & \frac{(\beta-1)v_2v_3}{\mp p\nu^2} & \frac{(\beta+1)(p+1)+2(\beta-1)\frac{v_3{}^2}{\mp\nu^2}-2}{2p} \end{bmatrix}$$

and

$$\mathbb{H}^-(\mathbf{R}') \cong \begin{bmatrix} \left(\frac{(\beta+1)(p-1)-2}{2p}\right)ct \pm \frac{1}{I}\beta v_1r_1 \pm \frac{1}{I}\beta v_2r_2 \pm \frac{1}{I}\beta v_3r_3 \\ 0 \\ 0 \\ 0 \\ 0 \\ \pm\frac{1}{I}\beta ctv_1 + \left(\frac{(\beta+1)(p+1)+2(\beta-1)\frac{v_1{}^2}{\mp\nu^2}-2}{2p}\right)r_1 + \left(\frac{(\beta-1)v_1v_2}{\mp pv^2}\right)r_2 + \left(\frac{(\beta-1)v_1v_3}{\mp pv^2}\right)r_3 \\ \pm\frac{1}{I}\beta ctv_1 + \left(\frac{(\beta-1)v_1v_2}{\mp pv^2}\right)r_1 + \left(\frac{(\beta+1)(p+1)+2(\beta-1)\frac{v_2{}^2}{\mp\nu^2}-2}{2p}\right)r_2 + \left(\frac{(\beta-1)v_2v_3}{\mp pv^2}\right)r_3 \\ \pm\frac{1}{I}\beta ctv_3 + \left(\frac{(\beta-1)v_1v_3}{\mp pv^2}\right)r_1 + \left(\frac{(\beta-1)v_2v_3}{\mp pv^2}\right)r_2 + \left(\frac{(\beta+1)(p+1)+2(\beta-1)\frac{v_3{}^2}{\mp\nu^2}-2}{2p}\right)r_3 \end{bmatrix}$$

is obtained.

Hence, the algebraic equation of Lorentz transformations is obtained from the desired matrix representation. Also in the transformation matrix in above for $p = I^2 = -1$, $c = 1$ we obtain following as

$$\mathbf{U} = \begin{bmatrix} \beta t & 0 & 0 & 0 & 0 & \mp i\beta v_1 & \mp i\beta v_2 & \mp i\beta_3 v_3 \\ 0 & \frac{-(\beta-1)v_1{}^2}{\mp\nu^2}+1 & \frac{-(\beta-1)v_1v_2}{\mp\nu^2} & \frac{-(\beta-1)v_1v_3}{\mp\nu^2} & \mp i\beta v_1 & 0 & 0 & 0 \\ 0 & \frac{-(\beta-1)v_1v_2}{\mp\nu^2} & \frac{-(\beta-1)v_2{}^2}{\mp\nu^2}+1 & \frac{-(\beta-1)v_2{}^2}{\mp\nu^2} & \mp i\beta v_2 & 0 & 0 & 0 \\ 0 & \frac{-(\beta-1)v_1v_3}{\mp\nu^2} & \frac{-(\beta-1)v_2v_3}{\mp\nu^2} & \frac{-(\beta-1)v_3{}^2}{\mp\nu^2}+1 & \mp i\beta v_3 & 0 & 0 & 0 \\ 0 & \pm i\beta v_1 & \pm i\beta v_2 & \pm i\beta v_3 & \beta & 0 & 0 & 0 \\ \mp i\beta v_1 & 0 & 0 & 0 & 0 & \frac{-(\beta-1)v_1{}^2}{\mp\nu^2}+1 & \frac{-(\beta-1)v_1v_2}{\mp\nu^2} & \frac{-(\beta-1)v_1v_3}{\mp\nu^2} \\ \mp i\beta v_2 & 0 & 0 & 0 & 0 & \frac{-(\beta-1)v_1v_2}{\mp\nu^2} & \frac{-(\beta-1)v_2{}^2}{\mp\nu^2}+1 & \frac{-(\beta-1)v_2v_3}{\nu^2} \\ \mp i\beta v_3 & 0 & 0 & 0 & 0 & \frac{-(\beta-1)v_1v_3}{\mp\nu^2} & \frac{-(\beta-1)v_2v_3}{\mp\nu^2} & \frac{-(\beta-1)v_3{}^2}{\mp\nu^2}+1 \end{bmatrix} .$$

Also, it is obtained by taking $< \vec{v}, \vec{v} >= -\nu^2$ by and a velocity $\vec{v} = v_1e_1 + v_2e_2 + v_3e_3$ in the reverse direction generally as follows:

$$\mathbf{U} = \begin{bmatrix} \beta & 0 & 0 & 0 & 0 & -i\beta v_1 & -i\beta v_2 & -i\beta_3 v_3 \\ 0 & \frac{(\beta-1){v_1}^2}{\nu^2}+1 & \frac{(\beta-1)v_1v_2}{\nu^2} & \frac{(\beta-1)v_1v_3}{\nu^2} & i\beta v_1 & 0 & 0 & 0 \\ 0 & \frac{(\beta-1)v_1v_2}{\nu^2} & \frac{(\beta-1){v_2}^2}{\nu^2}+1 & \frac{(\beta-1){v_2}^2}{\nu^2} & i\beta v_2 & 0 & 0 & 0 \\ 0 & \frac{(\beta-1)v_1v_3}{\nu^2} & \frac{(\beta-1)v_2v_3}{\nu^2} & \frac{(\beta-1){v_3}^2}{\nu^2}+1 & i\beta v_3 & 0 & 0 & 0 \\ 0 & i\beta v_1 & i\beta v_2 & i\beta v_3 & \beta & 0 & 0 & 0 \\ -i\beta v_1 & 0 & 0 & 0 & 0 & \frac{(\beta-1){v_1}^2}{\nu^2}+1 & \frac{(\beta-1)v_1v_2}{\nu^2} & \frac{(\beta-1)v_1v_3}{\nu^2} \\ -i\beta v_2 & 0 & 0 & 0 & 0 & \frac{(\beta-1)v_1v_2}{\nu^2} & \frac{-(\beta-1){v_2}^2}{\mp\nu^2}+1 & \frac{(\beta-1)v_2v_3}{\nu^2} \\ -i\beta v_3 & 0 & 0 & 0 & 0 & \frac{(\beta-1)v_1v_3}{\nu^2} & \frac{(\beta-1)v_2v_3}{\nu^2} & \frac{(\beta-1){v_3}^2}{\nu^2}+1 \end{bmatrix} [20].$$

5 Conclusions

In this study, the well-known Lorentz transformations in the physics were studied with elliptic biquaternions. Transformation matrices of 4×4 type consisting of elliptic components and 8×8 type consisting of real components were defined with the equality of the elliptic biquaternion $\mathbf{R} = ct + Ir$ combining space and time. The matrix equality 4.8 defining the relativistic transformation of 4−dimensional matrix expressions was given in 4.20. We also gave an easier and rather useful the expression of transformation $\mathbf{U}$ through the elliptic biquaternionic transformation matrix that we defined. Thanks to by expressing its matrix containing elliptic components in 8×8 type, we executed that mathematical expressions are obtained with fewer operations by making it possible to express the transformation of the relativistic transformation relation in the 8×8 type matrix representation in a simpler way. The recommended method may be preferred according to the type of operation, depending on the reader preference in order to be clear and simple. The form of expression in 4.1 and the physical results obtained with the help of 4.16 and 4.18 are provided easily interrelated. It is possible to express basic equations of electromagnetism with many physical quantities of different properties such as electric and magnetic fields, energy and momentum quantities, electrical current and charge density, scalar and vector potential which are closely related to each other through the equations and matrices obtained through this study, by means of elliptic biquaternion. Therefore, the expressions obtained are important in terms of being versatile and obtaining relativistic equations which very important in physics.

References

[1] B.L Van Der Waerden, *Discovery of quaternions*, Math. Magazine **49** (1976) 227–234.

[2] H. Hilmi Hacısalihoğlu, *Hareket geometrisi ve kuaterniyonlar teorisi*, Gazi Üniversitesi, Fen Edebiyat Fakültesi, Yayınlari (1983).

[3] W.R. Hamilton, *Lectures on Quaternions*, Hodges and Smith (1853).

[4] A. Kyrala, *Theoretical physics: applications of vectors, matrices, tensors and quaternions*, Philadelphia:Saunders Company (1967).

[5] L. Silberstein,*Quaternionic form of relativity*, Philosophical Magazine **23** (1912) 790–809.

[6] K. N. Srinivasa Rao, A. V. Gopala Rao and B. S. Narahari, *On the quaternion representation of the proper Lorentz group SO(3,1)*, Journal of Mathematical Physics **24** (1983) 1945–1954.

[7] R. Dahm, *Complex quaternions in spacetime symmetry and relativistic spin-flavor supermultiplets*, Physics of Atomic Nuclei **61** (1998) 1885–1891.

[8] K. Özdaş, *Fiziksel niceliklerin kuaterniyonlarla temsili*, Fen-Edebiyat Dergisi **2** (1939) 101–113.

[9] M. Tanışlı, K. Özdaş, *Application of quaternion representation to Stanford manipulator*, Balkan Physics Letters, **5** (1997) 65–68.

[10] J.P. Ward, *Quaternions and Cayley numbers: algebra and applications*, Kluwer Academic Publishers (1997).

[11] C. Manogue, J. Schray, *Finite lorentz transformations, automorphisms and divison algebras*, Journal of Mathematical Physics **34** (1993) 3746–3767.

[12] S. De Leo, *Lorentz group and Dirac equation*, Foundations of Physics Letters, **14** (2001) 37–50.

[13] S. Demir, *Elektromanyetizmanın kompleks kuaterniyonik dnm bantları*, Anadolu University Journal of Science and Technology **7** (2006) 247–253.

[14] Y. Tian, *Biquaternions and their complex matrix representations*, Contributions to Algebra and Geometry **54** (2013) 575–592.

[15] M. Jafari, *On the matrix algebra of complex quaternions*, TWMS Journal of Pure and Applied Mathematics DOI: 10.13140/RG.2.1.3565.2321 (2016).

[16] T. Y. Lam, *The algebraic theory of quadratic forms*, Massachusetts: Benjamin (1973).

[17] M. L. Mehta, *Matrix theory: Selected topics and useful results* Hindustan Publishing Corporation (1989).

[18] R. S. Pierce, *Associative algebras*, Springer-Verlag (1982).

[19] B. L. Van Der Waerden,*A history of algebra from al-Khwarizmi to Emmy Noether*, Springer-Verlag (1985).

[20] S. Demir, *Lorentz dönüşümlerin kompleks kuaternionlarla incelenmesi*, C.B.U. Fen Bilimleri Dergisi **3.1** (2007) 1–14.

[21] K. E. Özen, M. Tosun, *Further results for elliptic biquaternions*, Conference Proceeding of 7th International Eurasian Conference on Mathematical Sciences and Application **1** (2018) 20–27.

[22] Yaglom, *Complex numbers in geometry*, Academic Press (1968).

[23] A. A. Harkin and J. B. Harkin,*Geometry of generalized complex numbers*, Mathematics Magazine **77** (2004) 118–129.

[24] T. Erişir, M. A. Güngör, *Holditch-type theorem for non-linear points in generalized complex plane*, Universal Journal of Mathematics and Applications **1** (2018) 239–243.

[25] T. Erişir, M. A. Güngör, *On Fibonacci spinors*, International Journal of Geometric Methods in Modern Physics (2020) doi: 10.1142/S0219887820500656.

[26] K.E. Özen, M. Tosun, *p-Trigonometric approach to elliptic biquaternions*, Adv. Appl. Clifford Alg. **28** (2018) 3–16.

[27] K. E. Özen, M. Tosun, *elliptic matrix representations of elliptic biquaternions and their applications*, Int. Electron. J. Geom. **11** (2018) 96–103.

DOI 10.2478/9788395793882-[illegible]

TABLE of CONTENTS

Bei Fragen zur Produktsicherheit wenden Sie sich bitte an:
If you have any questions regarding product safety,
please contact:

Walter de Gruyter GmbH
Genthiner Straße 13
10785 Berlin
productsafety@degruyterbrill.com